Born to Croft

Ena MacDonald

Published by Linen Press, London 2024
8 Maltings Lodge
Corney Reach Way
London W4 2TT
www.linen-press.com
© Ena MacDonald 2024

A CIP catalogue record for this book is available from the British Library.

Cover Design: Lynn Michell
Photographs: Enhanced by Ged Yeates
Typeset by Zebedee
Printed and bound by Lightning Source
ISBN: 978-1-7394431-8-4

Dedicated with gratitude to the
readers and editors of Am Paipear.
Without their encouragement and support
this book would not have been possible.

About the author

Ena MacDonald was born in 1940 in Ardbhan, Kyles, North Uist and has spent almost her entire life in crofting. One of four sisters, and with one brother, she grew up on her father's croft on the machair – the coastal wild flower meadows found in the Hebrides. As a young woman, she left home and spent ten years working first in Glasgow and then in Australia. She returned in 1966 to help her father on the croft, doing everything from cutting peat to harvesting to growing and selling vegetables, and collecting and selling eggs to bring in an income. Quickly, it became her passion. She recalls creating a stir driving her father's little red tractor with her blonde hair in a ponytail and often with her young son, Angus, at her side. After her father's death, she took over the croft, played a full and active role in all crofting affairs and travelled widely as a representative of The Crofting Federation. She wrote occasional articles for the Scottish Farmer and Highland Cattle Society before her regular monthly articles for Am Paipear from 2006 to the present. In 2006 Ena was awarded the MBE for services to crofting and to the community in North Uist.

Note from the editor, Lynn Michell.

It has been my privilege and pleasure to work with Ena MacDonald on this fascinating record of her life in crofting. In these articles she recounts engaging, dramatic and humorous stories of crofting life and raises important questions about its past, present and future. She is a pioneering spirit, the loving matriarch of a large family, someone who truly understands and respects animals, and a woman who has achieved an astonishing amount at a time when crofting was more often left to the men. I have edited the articles with a light touch (I hope) and with respect for Ena's distinct, oral story-telling style.

My editorial assistant, Aurelia Knight, has edited this book with me, adding sensitive, perceptive comments and questions, and spotting the many typos I missed.

Lynn Michell

Note from Ena, March 2024.

Over the years, so many people have asked me, "Why don't you put those stories in a book?" In the late 70s, I wrote to the Highland Cattle Society and had articles published in their newsletters and journals as well as various other magazines. In 2003, I became chair of the Scottish Crofting Federation and decided that the Am Paipear – which was a local monthly paper – would definitely be most suitable for me to write about crofting matters. As crofting was my passion, the local crofters enjoyed reading the political and practical news. I had many treasured memories that were still so important to me and my stories were very varied. I enjoyed writing over the years and I do hope that you will enjoy reading this collection of my articles.

The People

Angus MacDonald	Ena's son
Michelle MacDonald	Ena's daughter-in-law, wife of Angus.
Ellie, Fraser, Sarah, Alexander	Ena's grandchildren
Carianne MacDonald	Fraser's wife
Flora	Ena's sister
Jessie	Ena's sister
Agnes	Ena's sister
Ewen	Ena's brother

2006

February
Moorland Grazing. A Learning Curve

In 1962, the crofters of Kyles Paible Township decided to fence 160 acres of moorland common on the slope of Marrival which hadn't been grazed since the 1920s. On it were ruins of shielings[1]. It was a dry summer and with the help of an old David Brown tractor the crofters fulfilled their dream of returning cattle to the moors. The heather had overgrown the thin grazing so the crofters were rather disappointed because the cattle, used to more green grass, were not impressed with their new surroundings. Although the cattle were slightly lean, they remained healthy. In the late 60s the plan to return the cattle to the moorland was abandoned.

In 1976, I thought if the shielings grazings were successful long ago, why not now? On a beautiful day in July we took our twelve cows, calves and yearlings, a mix of Highland and Shorthorn, on the five mile trek to Marrival and left them there for three weeks. Again, they weren't happy but they remained healthy. I was determined not to give up so the following March we burned the heather. That was nerve-racking because there is a small Forestry Commission forest along the east side of the fence and the wind was blowing from the east but, lo and behold, after half an hour, the wind changed to the west. Thankfully, with volunteers' help, we managed to save the forest and burn the 160 acres!

Summer 1978, the hill was green and the cattle were settled and thriving. The bird population benefitted and increased. We would often come across grouse nests of ten eggs and see interesting species like the golden plover, the sweet little dunlin, the long-beaked curlew and the very rare whimbrel. And the less welcome hen harriers and golden eagles. Struan Garbh, a semi-underground stream of pure crystal water, flowed down the side of the hill supporting an array of flowers and herbs.

To return the cattle and sheep to the moors would bring a multitude of benefits to wildlife and soil. Quality of meat is unsurpassed and the grazing on in-bye can be saved for winter. It would encourage shared herding and shepherding would increase the community spirit. It would be a treat for birds, bees, beetles, butterflies and invertebrates. The soil would get richer with the added dung. Calves born to dams that had grazed the previous summer on the moor never got white scour.

Bill Lindsay, a drover who used to buy eighteen-month-old cattle in Uist, told me, 'This winter they will be housed, next summer they will graze on the moor where they will grow but stay lean, next summer they will be on the best of grazing and if you could see them then, you wouldn't know them.' Those were the days when cattle rearing was a joy, without tags, paperwork and deadlines, and beef tasted as it should. Hopefully in 2007, a Land Management Contract menu will give realistic support to Common Moorland Grazing. The SCF[1] and other like-minded organisations are trying to secure this.

My family continued to use Marrival until 1997 when the fence was in bad repair. Maybe the policy makers will think again when they realise that biodiversity is disappearing.

OTMS/OCDS

Since 7 November 2005, all cattle born after 1 August 1996 can be slaughtered for human consumption. However, if they are over thirty months old when slaughtered, they must be tested for BSE[2]. Cattle born before 1/8/06 can enter the new OCDS[3] and compensation is paid to the producer. But, they must travel and, although not eaten, they must be fit for human consumption so any cattle that are lame or too old to travel are worthless and can be killed on the farm humanely by either a vet or gun-licensed person. These cattle are classed as fallen stock and although the Highlands and Islands have permission to bury stock, it is forbidden if they are born before 1/8/96.

Producers of slow-maturing cattle had looked forward to the ending of the ban on OTMS[4] entering the food chain, but relief turned to despair when producers approached their nearest abattoir to discover that only six in Scotland are licensed to slaughter OTM cattle. The problem for abattoirs is having to separate under-30 months and over-30 months until the BSE test is returned negative, and some licensed abattoirs cannot handle cattle with horns. If a cow born before 1/8/06 breaks a leg or has an accident which makes it subject to emergency slaughter, it is eligible for the OCDS.

BSE has been a thorn in my flesh for a long time and I've asked many questions and I've been given many answers, but I can never understand why such drastic measures are necessary. The £400 million a year spent on compensating farmers for having to incinerate cattle could have been put to much better use. As 93% of BSE cases were in dairy cattle, why were extensively reared beef cattle treated in the same manner? I'm sure most of you are aware that it is legal to do home kills. The age of the beast doesn't matter as long as it is slaughtered humanely and you must not sell any of it.

Crofting Foundation News

Not long ago, I listened on the radio to Morag MacLeod of Brue in Lewis, now well into her nineties. What an amazing woman. It was the most interesting interview I've heard for years.

I had the privilege of meeting Morag at the Barvas show last summer. She was the picture of health despite her age, her complexion and skin without blemish. She talked about crofting at the beginning of the 20th century when all the children helped with the daily chores, the milking, carrying peats home, carrying water from the well, a bucket in each hand. The soil, turned with a spade, was nurtured, producing potatoes, oats and barley. Picture a crowd of children planting potatoes, competing with each other to do the most work. Can you imagine what an artist's dream it would be?

Morag's talk made me think of how sadly crofting has changed. The Crofting Reform Bill states that agriculture doesn't have to be the prime industry of crofting. If a croft tenancy is for sale it won't necessarily go to the crofter next door but will go instead to the highest bidder and the highest bidder is usually only interested in selling house sites to those who can afford them. Does anyone have the answer? If the government can come up with millions of pounds for community buyouts, can they not come up with a substantial out-goer scheme?

Let's reflect on it. The imaginary canvas of Morag's childhood will fade if we don't put on a fresh coat of paint and get our children interested in a way of life which is quite unique. Let's all pick up a brush and cover the faded parts.

April

The AGM of SCF was held in Strathpeffer on 14th March. I have stood down as Chair but re-elected as Director for another year. Norman Leask from Shetland is our new Chairman, a very able crofter who rears sheep and cattle and has a passion for what he does. He has been a Director for three years and has represented our members in Edinburgh and Brussels. Hopefully he will visit Uist in June. The Vice Chair is Neil MacLeod from Stornoway. Neil joined the board three years ago too so you can be confident that the SCF[5] will continue to do what is best for its members. I am still Chair of the North Uist branch and we still have to hold the Uist and Barra Area AGM.

The College of Agriculture has been given instructions to carry out a feasibility study on a Uist meat processing community-owned business so let's hope it doesn't take too long. Now that the ban on exporting beef has been lifted, there should be plenty of demand for mature beef.

Do Highland Cattle have Table Manners?

In March 2005, a black yearling heifer was needing extra nourishment so was separated from the rest. Along with a similar bullock, she was fed oats in a bucket and the two of them spent two months together. When grazing became available, she returned to the others. We had been feeding the heifers bruised oats but the black one wouldn't touch them. One day I thought, surely you're not looking for a bucket? I had one and cunningly managed to get her away from the rest and offered it to her. I could hardly believe my eyes. She dived in and licked up the contents. So every day now she sneaks up beside me and gets her bucket out of sight of the others. It's obvious that cattle have good memories! Maybe next year she'll be wanting a knife and fork to cut her silage!

May

Imagine you were a goose. It must feel disheartening not to be welcomed everywhere. The numbers of geese are increasing every year and crops and grazings are ruined but even so, all we are allowed to do is 'manage' them. The remit of the NGMRG[6] is to protect crops and especially the indigenous seed grown on the Hebridean islands. SNH[7] pays for lethal scarers, mechanical (gas-gun) scarers and kites and that we appreciate. But this is a very short-term measure. In the last four years, the barnacle geese arrive in December and leave in April. About eight hundred seem to land together and they just love the short green grass on the in-bye[8]. Those who undersow grass-seed will only get half the crop expected and on a ten acre field, it is impossible to see a clean square metre of ground.

The cattle and sheep are trying to find grass but they cannot compete with the geese and their silage is also fouled. In April and May, stock is usually kept off the re-seeds to give it a chance to grow but the geese have already moved there with most of them nesting on the small islands in the lochs so they have free board and lodging. The cattle have to be fed for longer and the time for moving stock to graze the re-seeds therefore gets later each year.

The greylags stay all year round. In August, when the corn is ripening, they descend like locusts which is when the scarers are employed, but scaring just moves them from one township's machair[9] to another and to the re-seeds, which were earlier so ravaged by the barnacles. The estates have the shooting rights and crofters can get a licence, but how many crofters have the time to be shooting geese for long enough to make any difference?

The agri-environment schemes dictate that livestock must not graze on herb-rich ground from 1st May to 15th September. I can assure you that when 15th September comes, the herbs we are trying to protect are anything but palatable to stock. Is it not ironic to keep ground stock-free while hundreds of greylags are eating and fouling it? RSPB[10] are trying to protect the corn bunting and crofters would like to cooperate by harvesting a percentage traditionally, but it is a losing battle.

I have attended many goose meetings and everyone around the table knows that there is a problem. But how much do they really care? Do we wait until a way of life is gone and a fragile habitat is further damaged before action is taken? I believe the only solution involves pricking the eggs and shutting the closed season for shooting. There must be a major cull to bring the numbers down to what they were in the late 70s. Surely it would be kinder to goose, sheep, cattle and crofters to reduce the numbers. We could then live in harmony, the way nature intended.

Geese are delicious when plucked and roasted, a very healthy meal! If you get one during the close season, you might even get an egg too. And if you are lucky it might even be a golden one!

July
by Susy Macaulay
For Sale: 8 Highland Cows. Allow 16 years for delivery
Sixteen years ago, A Scottish and a German farmer shook hands on a deal that would send eight breeding Highland cows from Uist to Cologne. Days later, the guillotine came down on the deal because a ban was imposed on the export of live breeding cattle to Europe amid the mounting BSE crisis.

Instead of calling the whole thing off, the farmers kept the deal alive hoping, as the months turned into years, and the years turned into a decade and a half, that the ban would be lifted. This cross-border friendship and trust between two farming families finally bore fruit last month when eight heifers set off on their long journey to Germany, sixteen years after the sale.

This all began in 1990 when Axel Potthoff and his wife, Petra, were touring Scotland looking for good breeding stock to build up a herd of Highland cattle on a farm near Cologne. One of the farms they visited was Ardbhan[11], and it was the start of a relationship that would endure the seemingly endless knocks and blows of an industry beleaguered by BSE.

'I first saw Angus showing stock at Oban,' said Axel. 'He and his cattle really impressed me so I decided to visit Ardbhan and found Ena and Angus keeping their Highlanders in the most traditional and natural way with excellent results. I was keen to place an order, but then the EU ban came down and I was completely stuck.'

Expecting an early lifting of the ban, Ena and Angus offered to look after them at Ardbhan but it was soon clear that the ban wasn't going to be lifted any time soon.

'It was a terrible time for cattle breeders,' Ena explained.'One minute our calves were fetching £2000 at auction and then they were fetching only a few hundred. But by this time we had become friends with Axel and offered to look after his eight cows until the BSE carry-on was over.'

The years rolled by and Axel's cows had calves of their own. In his agreement, Axel would choose whether to keep the cows or calves, always keeping his number at eight.

'We agreed this was a reasonable and friendly agreement,' Axel said. 'But quite apart from that, we were firm friends and in close contact. When we met Angus he was a bachelor. Now he's married with four children. He and his family have battled through thick and thin to build up their Highland herd and keep it of such a high standard.'

The EU live export ban was lifted in May this year. Axel decided he wanted to buy a young bull from Ardbhan for breeding with his cows, before July, so the race was on to get the paperwork ready. For Ena, this meant weeks of tests, checks, paperwork and sleepless nights. 'It would have been easier to get them tickets for the World Cup,' she said. 'It was a white knuckle ride waiting for test results and the amount of paperwork made us boggle. Far more complicated than before the ban.'

For Ena it was an emotional moment when the cows finally left Oban on the start of their long journey. 'I know each and every one of our cows. I'm pleased they're going to Axel's farm and taking the bloodline over there, but it's still a wrench because I'm so fond of them all.'

Axel's cattle have won national prizes so the Uist cattle can expect serious cosseting in their new home. Their arrival was an emotional moment for Axel too who confessed to shedding a tear when they came out of their transporter safe and sound.

'Our Uist heifers are extra special when you think of the time and trouble to get them here,' Axel said. 'And we have cows delighted by the arrival of a new young Scottish bull. We look forward to seeing the Ardbhan bloodline being established over here.'

December
Memories of crofting in the 70s

In the 1970s, crofters were more self sufficient so at this time of year we were all busy collecting seaweed off the shore to fertilise the potato patches.

Two seaweed factories were operating and many cut seaweed or gathered tangle for extra income. We used to hire the Big Grab from Alginate to load the trailers before the next high tide would take it away.

There have been huge changes in cattle and sheep rearing. Cattle used to be housed all winter. The cows were milked and their calves hand reared which left the calf very tame and easily handled. However, when Uist Calf Producers got their scheme going, there was good return for a suckled calf so the calves were left with their dams and most crofters kept the best milker as a house cow. Gradually crofters started out wintering their cattle.

In the late 70s, the original fences were needing replaced and sheep were running riot so in 1982 with open arms we welcomed a 5 year Integrated Development Programme, and for the first time ever, money was thrown at us. 86% grants for fencing, 60% for machinery and steadings, money for heifer retentions and lamb premium. However, tempting as the 60% was, the full-time crofter still had to find the other 40%. If he had a professional job, he could easily afford a new tractor and machines. Sadly the scheme budget was spent faster than was anticipated. The policy-makers were short sighted with emphasis on production but never a thought to the damage to the environment. We were forced to use artificial fertiliser and weed control, chemicals detrimental to our rare birds. Then conservation bodies woke up and decided they could teach us how to croft in environmentally friendly ways, so then came the Agricultural Environment Schemes which, you might say, went back to basics. BUT, instead of paying crofters to crop and manage their land as they did before, policy makers invented 'menus' restricting grazing on wetlands and on herb rich ground to certain times of the year. Those herb rich areas would never have been herb rich if they hadn't been looked after properly in the first place. I know a field which was grazed by cattle four times within the grazing period, and still it grew a magnificent carpet of wildflowers. Now it is choked with silverweed and trampled by geese. The same applies to wetlands. There are too many rules and regulations – and why does it take so long for the policy makers to understand the harm being done?

Stop this waste! In the 40s and 50s we didn't buy feeding stuff from the mainland and not a blade of grass was wasted and we had more flowers, fauna and wild birds. Children are not getting involved in croft work nor are they taught where their food comes from. The opportunity isn't there for them because now everything is mechanised.

I know there has to be progress and I know the climate is changing. Not even the policy makers can change that.

2007

April

If you've never handled cattle, you might walk past a pen and imagine they've gone in like a dog goes into a kennel. Unfortunately, it isn't that simple because cattle are very intelligent and always on the defensive – and they don't like vets sticking needles into them, so they don't willingly walk into pens.

At the end of January it was necessary for us to do some blood tests on the stirks[1] getting sold off the island. The house cow I milked was the boss of this mixed lot so for about three weeks I left all the pen gates open, and after milking her, I would put portions of sugar beet in the pens and walk away. The cattle were familiar with the routine and were convinced nothing would happen to them when the gates were open.

Vet day arrived and as usual the cattle went in fearlessly, but then the gates were shut and the nasty needles were inserted. The blood samples were posted the next morning to the Edinburgh laboratory. The following Monday I phoned for results and I was flabbergasted when told they hadn't arrived. Every morning for a week I phoned to be told the same.

The only option was to retest the cattle. Thursday 15 February was a horrendous wet and windy day. Normally I wouldn't be out except to milk the house cow but since the cattle had experienced the sharp needles two weeks previously, the only way they would go into the pens would be to follow the house cow which they did, and Angus and Angus Og[2] shut the gates behind them.

The wind speed was about 60 mph and I had done all that I was capable of doing so was about to make my exit. I felt the wind pushing me and at that moment the pet bullock, the one that had injured his leg in a cattle grid, moved backwards to get away from the cheeky heifer that was stealing his sugar beet, and the force of the wind and him knocked me off my feet and I came down heavily on the concrete floor. The cattle sensed something was wrong. They could have trampled me but they moved away and the men carried me into the house.

I was flown by air ambulance to the Southern General Hospital where I got a hip replacement. I was there for four days and the surgeons were first class. It was nice to see my two sisters and Uist friends, although I wasn't feeling very sociable. Ruth and Nev Thomas drove all the way from Wales to visit me – imagine! I was then flown by air ambulance to the Uist & Barra Hospital.

That's why I started writing this article. I couldn't hold back the tears when I was wheeled into an immaculately clean ward and surrounded by familiar faces. I just don't know how to thank the staff. They absolutely spoiled me and my strength returned in leaps and bounds. No wonder the island girls nursing in the city hospitals are so highly respected. The food was good but, I must be honest, not the ice cream. I was overwhelmed by the kindness shown to me. I want to thank all who visited me and all who sent me cards, flowers, chocs, and the cleaners who diligently made the ward look like home. I want to thank the physiotherapist who got me back on my feet and advised me what movements I should avoid. My most special visitor was my dear friend, Mia MacCorquodale from Claddach Kyles. I was absolutely thrilled when she walked into the ward.

Now I'm convalescing with Angus and Michelle, and the children are a tonic. Alexander, the one and a half year old, has me hooked on Bob the Builder, and having so much time on my hands, I can spend hours with him. Michelle has even been cooking spinach which I thought I didn't like. I shall soon be back to normal.

Going back to the cattle blood samples, five days after we did the second test, the results from the first blood samples arrived, although they had denied receiving them. The second test and the gathering of the cattle had been unnecessary.

June

The Scottish Crofting Foundation has been nominated for a prestigious award, Campaign of the Year, put forward by SCVO[3]. It was shortlisted because of the tireless campaigning of volunteers and staff against an unwelcome Crofting Bill drafted by civil servants. Their annual conference will be held in Dingwall and the theme will be Crofting Culture. Dingwall is within easy travel distance so I hope some of our Uist and Barra members will attend. The Crofters Commission has appointed Drew Ratter as its new Chairman and three new board members, one of which is Angus McHattie who used to work for the SCF. I believe the Commission is going to address the problem of underused croft land.

During the recent warm spell, I was thinking back to my school days. During lunchtime, we would go for walks looking for birds' nests and early wildflowers. For our parents, spring was always a lean time but a very busy one – peats to be cut, ploughing and planting to be done, dung heaps to be cleaned and spread on the land – and all without machinery!

The diet was simple – salted meat and fish – as very few animals were fit for slaughter. The potatoes would be running low which meant more scone baking for the women. You couldn't buy vegetables and the only greens available were nettles. Cows calved in May so there was milk available and it was a good time for hens and ducks to be laying. Broody hens were given a dozen eggs and nothing was tastier than a young cockerel cooked in homemade butter. People living near the sea could collect shellfish and we used to sail to Kirkibost Island to gather seagulls' eggs. The yolks were orange. We always left one or two in the nest and if there was only one egg, we didn't touch it. Those eggs made lovely pancakes. I don't think I could enjoy one now though because they are rather strong.

In those days you could take a few dozen eggs to the shop and get other messages. A friend of mine had to carry a tin pail full of eggs, only she fell and most of the eggs broke and she arrived at the shop crying. She had a list of messages to get but the shopkeeper, Flora MacLean, took pity on her, cleaned up her pail and gave her all that she needed. Can you imagine walking into a supermarket today without any money and a pail of broken eggs!

July

Childhood memories: At the Peats

I'm looking out of my kitchen window and can see the black round rock appear above the water. I would still be unable to paddle across to the other side of the ford.

When I was a child, the tides were at different levels and when this rock surfaced above the water, we left the house as we could safely paddle across. During the war, when most of the men were away, the women carried on with the croft work so my mother, like most mothers, had to take the children with her to every job.

One day she was going to cross the bay with the five of us. She left a few minutes too early, pushing the pram with me in it and the other four were holding hands across the outgoing tide. My sister, Agnes, a four year old, started to panic and wouldn't move a step. Poor mother couldn't let the pram go or it would float away so she screamed at my brother, Ewen, to grab and drag his sister across. He seemed to be rooted to the seabed and it took him what seemed like hours to realise that he was able to help. Had she gone too far to the right, the sea was deeper and the current would have lifted her. If we went around the bay it would be an extra two and a half miles. I think that day my mother stopped at Sine Alasdair's Millburn Cottage for a relaxing cup of tea. Another mile and she would arrive at the peat bogs where she would work hard and fast with the help of the four little ones and leave in time to catch the incoming tide.

During the late 40s and 50s, our father would take us all in the horse cart. The bottom of the cart would be full of sand sods to repair the peat road, and we would have a good supply of scones spread with homemade butter, cheese, crowdie[4] or eggs, often hard-boiled.

The kettle stayed outside all summer and whoever filled it at the river would have to rinse out the spiders and beetles. I'll never forget the strange taste of tea made with pure peat water, no chloride then, nor did I ever hear of bugs! Father would always light the fire and make the tea, but he was never domesticated enough to make tea in the house. We all loved tea time, provided the midges and clegs[5] stayed away. During our break, we would sometimes go down to the loch to get water lilies. Then back to loading the cart with peats at the roadside and the last load of the day would be topped up and taken home. If the tide was coming in, we would go further and cross at Horisary where the water was less deep.

Peat work was in seven stages. After the winter rains, the drains had to be unblocked to let the area dry better. Then there was turfing (*feanadh*), removing about four inches deep, eighteen inches wide and one hundred and fifty yards long with a spade. Next was cutting (*buain*) with the peat iron (*troisgair*), one person cutting, another throwing it up onto the bank. Depending on the weather, usually after ten days, lots of six peats (*ruthain*) were stood up and put to dry. After another ten days they were put into bigger heaps (*cruadhachadh*), then to the roadside and home. There were no plastic bags so we used a wooden hand barrow to move them to be assembled for the cart. On the way home on the Committee Road, we always stopped at the *fuaran*[6], a spring west of the cattle grid, to have a refreshing drink.

During the war years, there wouldn't be enough peat and my mother and the older children would gather every bit of driftwood found on the shore to saw and fit into the Modern Mistress stove. When I think of the hard work involved! That's why I resent the November bonfires and sticks being imported on to the islands.

The peat machines have made a huge difference, and with oil and electricity we don't see so many peat stacks. However, many people love the scent of peat smoke and sit by the light of the peat fire flame.

Sorry But Cows Can't Fly

Once upon a time it was a joy to herd the cattle from one pasture to another. They could move at their own pace, stop at the roadside to drink or graze or even rest. Now it's a dread, mainly because of impatient, thoughtless road users. If the cows can be moved before peak traffic it's less stressful, but we depend on tidal waters and the spring tide is always out in the middle of the day.

There are wonderfully helpful people on our roads so it's a shame that the ignorant, selfish minority have to blur this. There was a lorry driver who grasped the situation and stayed well back, turned off his engine and acted like a barrier and the beast went where it was supposed to go. On another occasion there was a van driver who drove ahead of the runaway heifers and managed to turn them around.

A very small minority will give you abuse. Recently three black ATVs drove past us on the Committee Road[7] without slowing down and the cattle were almost flying to get off the road. It would help if oncoming traffic would stop in a passing place or reverse back because stopping in the middle of the road frightens the animals and if there's a ditch, where can they put their hooves? That's when wing mirrors can be broken. I have seen people approach us on Ard Heisker Bridge, almost unbelievable.

Using a house road end as a passing place is good but not if you stop opposite the entrance because you are encouraging the cattle to walk up the driveway. It is always harder to get past if you're in a car behind them. Please let us know if it's an emergency and maybe something can be done. If you are just rushing home for your dinner or favourite programme, cattle traffic delays don't happen often, and you tolerate road works. Maybe it has come to the sad day when we need a police escort.

Mother told me of hard times when people really valued meat. She and my sister, Flora, would walk the twelve miles from Kyles to Trumisgarry and back again, bringing a fat cast cow on a rope to be slaughtered. This beast would be shared by four families. I'm sure there wasn't any traffic on the road that day. Nobody sponsored their walk and there wasn't a car to give them a lift there.

September

Life Nature is a dedicated EU fund to conserve highly important habitats. At a recent meeting organised by RSPB and held in Inverness we heard about the Machair Funding Opportunity and we hope to send in an application by the end of November. All organisations interested in conservation were invited. Becky Shaw represented the Scottish Crofting Foundation. Iain MacKenzie, from the RSPB, will be working full time on this and I sincerely hope that we can get a bid in.

SNH has been worried about machair conditions and the disappearance of some species of birds, flowers, fauna and invertebrates so the application has to convince the EU that funding will benefit and increase those species. The fund is for all Scottish machairs, but ours is quite unique. This funding is long overdue and it angers me that 'experts' keep asking the same questions such as why is less traditional harvesting being done and why are there fewer corn buntings?

If policy makers had supported schemes to encourage seaweed application and moorland grazing, and to reduce the greylag goose population and to support traditional harvesting, everything else would fall into place.

Recently, I attended a ministerial meeting at Nunton Steadings. Our new Environmental Minister, Mike Russell MSP, is getting to know the problems and views of the people living in the rural areas he represents. With him was the Chief Agricultural Officer, who is the Chairman of the National Goose Group, the Chief Executive of SNH, Chief Executive of Crofters Commission, and SEPA. Ivan Macdonald asked if and when there would be a greylag goose cull. There weren't many crofters there but enough to raise the temperature in the Steadings. It was obvious that SNH's Chief, Iain Jardine, knew little about our problems, but the short crash course at Nunton will keep him thinking for a while! It was agreed that SNH would decide on a viable population, not that we have any guarantee that something will be done if the numbers are too high. Of course we already know what our islands can tolerate and we don't need science or computers to tell us. Anyway, Mike Russell promised to follow it up and anything is worth trying.

Priority is given to protecting arable crops, but if they can't protect our summer grazings, we can't keep stock, so there won't be any corn. When that happens, goodbye corncrake, buntings and bumble bees because the geese will multiply. Maybe we can make a living selling goose eggs in baskets made from rushes! RSPB and SNH will then look around and employ project workers to try to discover why the Uist & Barra landscape has changed.

October

In September last year, one of my hens appeared with twelve tiny chickens and miraculously they all survived. With predators around and windy weather, it's a lot of work rearing and protecting them. They've been laying since February.

A month ago one of those chickens disappeared. I looked everywhere for her and when she didn't appear, I believed that a ferret had got her. Then about two weeks later, she appeared early one morning, very hungry and clucking so I know she must have been sitting on eggs somewhere. Then a few days later, exactly twenty-one days after her disappearance, I heard a cheep, cheep coming from behind some gates and there was my young hen with thirteen chickens! Beautiful balls of fluff they are and I wonder how they can all hide under the one hen!

I got the cat box and put it beside her and sprinkled some oatmeal inside it. Some say a hen is brainless but I disagree because this young hen walked into the cat box and her chickens followed. So then she was easily carried to the wooden chicken house which is safe from vermin.

Most chickens are hatched in June or July. Has it something to do with climate change? I don't think so. When we were young, cattle were herded and none of the houses were fenced, so there wasn't any long grass for the hens to hide in. And with more people around it wasn't easy for a hen to find a hiding place. If a hen started clucking and wasn't needed to rear chickens, she was put under a box for three days, usually without food. This was to make her very uncomfortable and it stopped her clucking or brooding, so when she became free, she started laying.

In those days, like all home produce, eggs were precious and there were no mainland eggs in the shops. I just fed the hen and her chickens with coarse oatmeal in a steel plate full of water. I put stones inside it to keep it firm and so the chickens didn't get wet. They can even stand on the small stones! It is very important for hens to have access to grass, so every day I cut a fresh sand sod of short soft grass. After a few days they can go into a vermin-proof pen outside. By the way, if anyone with children would like to come and see the chickens while they're still at the cute stage, I would very much like to show them off!

Alexander, my two year old grandson, loves all the feathered creatures. He calls them 'cok cok' or 'bik bik. He calls the cockerel 'oodle-oodle-oo'. They are worth every effort when I see his face light up from the pleasure he gets from watching them.

2008

February

Sitting here in front of the peat burning Rayburn, with a 70mph gale outside, I was thinking how different it was in the mid 1950s. Before dusk on a similar kind of day, my father's voice would be urgent: 'If you need anything outside, get it now.' Then he'd remind us to keep all the doors shut. I still follow his good advice.

Those were the pre-silage days with most cattle in-wintered. In January, we would be feeding the first corn stack with another five firmly secured in the stack yard. Every crofter's mind was fixed on the stacks. We didn't have discarded fishing nets to secure them, only hessian ropes, which we called *caitha*. If we lost our stacks to the wind, there was no feeding for the livestock. You couldn't go and buy hay from the mainland. When the stacks were first built, sheaves were put neatly in the barn so that corn was used for feed first. When the barn corn was nearly eaten, usually on a Saturday with the children at home, the ropes came off the stack and the barn refilled and then the stack would be secured again. The iron rims that surrounded the cart wheels would be thrown on top of the opened stack making it less vulnerable to the wind. I loved helping to feed the cattle and horses. The corn was usually oats and rye and you got to know each animal's appetite. The short cut sheaves with plenty grass below the band went to the weaned calves. The older cattle and horses could cope with a mix of oats and rye, but straight rye wasn't very palatable and the straw stem was coarse. The barley straw, grown on richer soil, was softer and contained more grass. The seed itself was fed to the hens that were kept in large numbers on every croft.

Before my time, the barley was milled and used for human food like scones and porridge. As soon as daylight came – we didn't have electricity until 1968 – the cattle were fed about three sheaves of corn and at midday were untied and let out for a drink of water. If the day was pleasant, they stayed for a few hours because, after being tied in a stall, they needed to exercise. When we had hard frost we left them inside, but buckets of water had to be carried to them. We got more frost and snow in those days and the temperature was much lower in winter.

The cows usually calved in May but as everyone milked their own cows, they would go dry early in the year but one with high milk yield was usually left so that we had milk right through the winter. I loved the winter crowdie. Milk doesn't go sour in winter so rennet had to be added and the crowdie tasted different. The cows calved in the byre, like dairy cattle. Cows that are tied should never be left alone whilst calving. My sisters used to tell me how they would sit in the byre and watch for the cow to go into labour. Once she calved, the calf was dried with straw and put into a corner in the barn. The cow would get a bucket of warm water with oatmeal and the best sheaves. The colostrum was milked and fed to the calf. Some people made colostrum cheese, highly nutritious, but I didn't like it. The calf was taken away from the mother at birth and the cow would smell the calf on our hands and coat and lick. Families were much bigger then and milk was precious, so they couldn't risk letting the cow see her calf in case she refused to be milked. I never liked that part of crofting. I felt so sorry for mother and calf. Systems have changed now and all the milk is imported from the mainland.

It's good to know that some crofters still continue to feed corn in the sheaf as I believe you couldn't get a healthier feed. The rest of us use silage which is fed by machine.

May

Most of you will have read the Committee of Enquiry's key recommendations for crofting. There will be a lot of debate and discussion before this becomes law so if you have an interest in the future, think hard and make your views known. One of the biggest issues is the recommendation to abolish the Crofters Commission. They have often let us down, but better the devil you know than the devil you don't. Would Crofting Boards be better? I find it difficult to decide what would be best. The good thing about this inquiry is that it has come from crofters. I want to see crofting continuing with agriculture given priority. With fuel prices soaring, we're seeing crofters going back to basics, peat cutting and growing vegetables, especially potatoes.

This has been the driest May I can remember. What a contrast to 1986 when we didn't have one dry day for the whole month. Fortunately the ground was full of water after the wet winter so the heavy soil is producing, but the machairs are struggling. On the other hand, it is pleasant for man and beast to feel the sun on their backs, and lambing and calving have benefitted. There are more birds too, and bird watchers all over the place. The corncrakes are quite visible. The bird songs combined with the corncrakes is like a choir with alto and soprano. Sadly, as usual, the geese are causing havoc and the deer are getting more cheeky so I'm afraid there's a lot to be sorted out.

I had two broods of chickens, eleven in each. One hen was lying on rotten eggs so I threw them away and wrecked the nest. I was still missing the hen and then found her sharing a nest with another. I've never seen this before. You would think they would fight. The chickens hatched, so two hens are sharing eleven chicks. I'm always over-protective and keep them in pens for safety but this pair took their chickens outside, so they can get on with it. Today my cat was eating their breakfast of potatoes and oatmeal with two broody hens next to her. I don't think she would touch the chickens. I have to put stones inside the water dishes in case the chickens get drowned. It's lovely rearing your own chickens but it's time-consuming. I guess I'm a glutton for punishment!

July

I have family members living in Paisley and on the occasions when I've travelled there, I spend the 45 minute journey admiring farms close to the railway line. Along the Ayrshire coast, I would be envious of the dry land that stretched for miles and could have provided winter shelter for all the cattle on Uist. Strangely I didn't see any livestock, and after a few more journeys I realised it was a golf course. As someone who saw no pleasure in a game of golf. I was horrified that livestock was excluded from the vast expanse of wintering ground.

Last November was my first visit to Askernish machair in order to report back to our SCF Head Office. This was similar ground to the Ayrshire coast and again there were no cattle or sheep in sight. I have experienced hard times rearing livestock, times when every blade of grass and sheaf of corn was precious. I couldn't believe that nearly all the main grazing areas here had been mowed, leaving the least nutritious grazing for the livestock. In the 70s the first fairways were 20 metres wide, but now, as well as the extension of a nine hole to an eighteen hole course, the fairways are 52 metres wide. The sad thing is that all this was done without the consent of the majority of shareholders because Storas Uibhist believe they have a right. That right to play golf was just given to Lady Cathcart, free of rent. South Uist Estate own this land but it is under crofting tenure.That gives the eleven shareholders the right to manage and protect their common grazings as they wish under the leadership of their grazings committee. Askernish crofters have been condemned for being defiant. I think they've been extremely polite and patient. Would other townships allow this to happen? I think not. In North Uist there would be strong and total opposition. You would imagine a community landlord would have crofters' rights as a priority, or are they trying to diminish those hard won rights?

Looking at it from another angle, it benefits the wider community and brings much needed revenue, but could it not be done on a smaller scale? The Grazings Committee did offer an area to accommodate an eighteen hole golf course located on the existing course, extending where necessary to the south and north outwith the dune system. This would have provided a protected grazing area where livestock would be undisturbed. The current plan has an impact on the whole of the grazing area. Excuse my ignorance but couldn't fairways be limited to a width of 20 m? Would they not be challenging enough?

Storas Uibhist maintains that golfing has improved the grazing but if that grazing is cut every week, what is left to graze? Already crofters are being reprimanded about where they allow their cattle to be fed. Will Askernish end up a replica of the Ayrshire coast?

August

We invited Bridget England to the North Uist branch of the Scottish Crofting Foundation AGM on 10th March. Bridget is doing advisory work for RSPB on the Scottish Rural Development Programme. Information and applications have to be done online but we agreed that Bridget could supply crofters with information on paper to be left in local shops to take home. There are a thousand pages and I fail to see why they can't be printed, or a copy left for browsing in the local Agricultural or College Office. This is supposed to be progress. They forget or don't care that many of us don't own computers.

In the March issue of Am Paipear[1] I read the riding stables' SOS. I hope that with public support, they will be able to continue. Horses are wonderful animals to work with and children gain a lot by learning how to look after them.

No doubt many of my age will be thinking back to the days when we used to work with horse and cart, preparing the ground for ploughing. I loved coming home from school, having dinner or tea, and then off to the machair. My sisters would have other jobs, but I was the tomboy and did most of the harrowing. My brother and father would have been ploughing and planting seed all day so when I arrived, the horses would be harnessed to the harrows, my father and brother would go home, and I harrowed until I covered and levelled down all that had been ploughed.

Of course the highlight of the job was unharnessing, finding a rock to stand on and jumping on the back of old Sally with young Sally walking behind us while I held her reins. One day when my brother was with me, I begged him to let me ride young Sally while he took the mother. Sally Og[2] was just about four years old. Ewen lifted me up – my first time on her back – but before I could tighten the reins she took off while I desperately held her mane. Of course it was bare back and she was keen to get back to the stable for the night, about one and a half miles away. Despite seeing cowboy and indian films on the mobile screens at school and envying John Wayne, it was frightening when it happened for real. As we got closer to the village, I could see the men folk standing outside probably waiting for me to fall off. Fortunately the stable door was shut and Sally had to put her brakes on when we reached it. I was full of childish pride that I managed to stay on. I don't know who got the worst row, probably poor Ewen for allowing me on Sally Og's back.

I laughed at Lachlan MacInnes's article suggesting I pull a plough with Highland cows. Actually Lachie, I have thought about it but it never went any further!

October

On the 6th October, a wild, wet morning, I travelled to beautiful Barra to attend the Scottish Crofting Federation Conference. The theme was Scottish Crofting Produce with interesting speakers on Direct Marketing, Good Food, Quality Stock and Quality Meat.

Tom Gray, a regular writer in the farming press, gave a most interesting talk on Highland stock and said how cattle from the Highlands and Islands do well when sold to the east. He emphasised the importance of a strict high health status and how careful we should be about importing stock. Gavin Jones, who organised our 2006 International Conference, gave his usual passionate talk about Selling High Nature Value. He is knowledgeable about our machairs. We had lively discussions with Mark Shucksmith about the crofting enquiry and with Michael Russell on everything from the Crofters Commission to bulls, geese and house burdens. Our MP, Angus Brendan MacNeil gave us a lot of support. I would like to thank everyone who made this gathering so enjoyable.

Every day we hear about the credit squeeze but if people ate more sensibly they would be able to tighten their belts. Why spend money on convenience meals when it's simple to cook from the cheaper cuts of meat. Use your biggest pot, half-fill with boiling water and salt, add 2-3 kg of mutton or beef and after boiling for an hour, add chunks of carrot and turnip and whole small onions or halved large ones. Simmer until the meat is tender and the vegetables are cooked. If there is too much fat, pour the stock into a large bowl and leave overnight. Next day, lift off the surplus fat, put the stock back into the pot, boil and add your favourite cereal – barley, lentils and diced vegetables – but only a few because the juice from the first cooking is still in the stock. Boil until vegetables are cooked and there's the tastiest broth for the cold weather.

You can cook the same way with birds. Geese are delicious. Rabbits seem to be free of myxomatosis and are a delicacy. Young ones can be pot roasted in butter and older ones can be stewed. Our family ate them in many ways. The boiled ones would be fried in dripping, or served covered with either curry or parsley sauce. Stock was never poured out but always made into soup. Potatoes were served with everything, mainly cooked in their jackets. Any potato leftover would be peeled and fried with an egg for breakfast.

We never ate pasta except macaroni. I still remember the delicious macaroni and mashed potatoes that May MacIntosh dished up for dinner in the school canteen, huge trays sprinkled with cheese and browned in the oven.

In Liniclate School, crofting is being taught to S3. Maybe the preparation and cooking of traditional meals could also be taught before those skills disappear. Too much ends up in the bin and it's time to go back to basics.

November

The Scottish Crofting Foundation invited Iain MacKenzie to explain what the Life EU Machair Funding Opportunity is.

Applications must be in by the end of November 2007 to allow a 2009 start so I'm disappointed that RSPB is delaying until July 2008 which means a 2010 start. In monitoring our machairs, SNH has found a decrease in some of our rare species of invertebrates, birds and wildflowers. Why treat this urgent issue so casually and why is only one person employed to put in a bid when time is limited? During our discussions, more enthusiasm was shown in South Uist for traditional crofting and the use of seaweed. Everyone agreed that geese are our biggest threat but while this project would bring more money for goose management, it would be for lethal scaring. And if this project doesn't start until 2010, it will be a useless exercise.

I was not in favour of the sheep welfare scheme. I believe it's morally wrong to rear lambs and then incinerate them. Another way can be found.

December

I have been invited to the launch of the New National Crofting Course at Sgoil Lionacleit. Michael Russell, MSP, Minister for the Environment, will be there and I hope I have the opportunity to ask him vital questions. As you all know, the bull scheme has come to an end and while the Government talks about an alternative, to my knowledge there isn't one, nor any substitute for a bull unless they've invented a laser beam that will get our cows pregnant!

The SCF has been desperately trying to get an increase in LFASS[3] payments. It was designated for less favoured areas yet we on the west coast get less than the fertile east. The Royal Society of Edinburgh has produced a report on the Future of Scotland's Hills and Islands with a debate in the Council Chambers in Stornoway in December. I intend to be there or maybe take part by video link.

Maybe it's a sign of old age, but I am getting tired of reports, consultations, debates, lots of words but no action. We need more support and fewer regulations. Why can't they deal with the problems caused by unfair rate of LFASS, cross checks, the thirty month beef rules, geese, deer restricting grazing on wet and herb-rich lands, very little payments for moorland grazing and seaweed application.

The Blue Tongue[4] petition against compulsory vaccination is gaining momentum. Instead of threatening us with a £5000 fine, the Government should fine the English farmers who import infected cattle from France. We must all be careful when importing stock to our islands.

2009

March
Scottish Crofting Foundation News
Sheep Tagging

As we go to press, Norman Leask, our past chairman, is meeting European officials on a farm near Pitlochry to discuss the electronic tagging of sheep. If this goes ahead, it will be a nightmare, especially for large flocks. And again it will involve more paperwork.

Bull Hire Scheme

We've won the *Battle of the Bulls* but the war is not over. We have been guaranteed another year. The government admits that the scheme was not breaking state rules, but why did it take so long to discover that? Despite the good news, it leaves cattle producers uncertain of their future. Money spent on meetings to discuss this scheme could have been put towards the overwintering of many bulls! Every support scheme becomes a cat and mouse game and despite modern technology, everything gets more complicated.

Blue Tongue

In January I attended a meeting in Edinburgh to try and make a case for voluntary vaccination. I presented the petition signed by most crofters and the Chief Government Vet, Charles Milne, was very sympathetic. The biggest stumbling block was that our local authority would need a regulatory order and a by-law in place, as is the case in the Shetland Islands. In Shetland, all imported stock is declared, they are examined and tested by a vet, and immediately afterwards transported to the farm or croft. They are then kept separate from the other livestock until the test results are in.

The vaccination period has been extended but pertains only to cattle producers on uninhabited islands so will only benefit crofters with sheep on very remote ground. Crofters with sheep on moors, hills and rough terrain can apply to their vet for an extension on welfare grounds. It's a pity that cattle aren't in this category, especially when they have to be vaccinated twice. I feel quite confident that a voluntary scheme can be in place by next year. Over 800 crofters signed the petition and that makes a strong case.

Geese

A research project could be underway in Uist during this 2009 breeding season. RSPB and SNH have agreed to test the effectiveness of egg pricking and oiling. If this is effective, at the Government's discretion, licences could be granted to crofters to follow on. However, we need to press for more shooting too. This mild spell is allowing some grass growth but when 800 barnacle geese are on a six acre field, what's going to be left for the livestock?

July

At the end of March, we sold forty-two heifers to the Forestry Commission. They wanted hardy, extensively reared stock used to hill ground and immune to ticks. It was great to see them go to the same home. Cattle have feelings too and sometimes their loyalty to one another should drive us to shame. So far, they are doing very well at Loch Katrine and we will keep in touch with their progress.

In the same week, seventy-eight bullocks were sold to United Auctions, most going to Aberdeenshire. Handling the bullocks in different fields was a nightmare as was driving them on the main road. I honestly can't understand the mentality of some drivers – I know the cattle don't pay road tax, but then again, how are they supposed to travel? So please, drivers, have some consideration for livestock. Mind you, last week we walked cattle from Vallay Island to Griminish and we met only patient and considerate drivers. One van driver actually turned back and stayed on the side road used by Griminish Pier fishermen and when the cattle approached, and the usual contrary cow decided to turn right, he got out of his van and chased her! So there are

Good Samaritans around. I hope he reads this because we want to thank him. He had a Lewis accent.

Back to the bullocks. They are finally boarded and have their passports ready. Every animal needs a passport with their date of birth, tag number, number of sire, name and address of owner – I'm not joking! The two livestock lorries were due at 8.30 am. I was dreading the loading because of the strength needed to load six bullocks into a normal sized trailer, but I didn't know Norman MacAskill and Iain McEachern were so organised and skillful. They reversed those huge vehicles towards our pens with just an inch to spare. The near ramp was opened and five bullocks were filtered into the near pen, and walked in like old cows going into a byre. The gate was shut behind them and then the next five went in. The smaller ones went in the top deck. There was no shouting, no sticks flying and I watched it all with enjoyment. Maybe the bullocks knew that they were off to lush green grass up TO their bellies. With the top deck full, a button was pushed and up went the ramp, opening the door for the lower floor. I could not admire those two capable hauliers more. We don't appreciate skills like theirs until we are involved in the operation. I hate to think where the biggest bullocks are now, but I'm sure there have been some tasty meals served up in Aberdeen.

August
Draft Crofting Reform Bill Consultation Paper 2009
I have had numerous requests from crofters concerning this important paper, so I thought I could help if I printed a summary of my personal response:

1. I strongly disagree with establishing Area Committees. The Crofters Commission should remain. The Board of Commissioners should be elected by crofters with maybe two by the Scottish Government. The Chairman should be elected by the Board. The Assessor Network should be strengthened and given training if necessary. The assessors advise and the Commission decides.

2. Using croft tenancy as Standard Security should be removed from the bill. It's a dangerous road to take. If a crofter, especially a young entrant, defaults, he can lose his croft before he even gets started and his Security of Tenure would cease. Equally dangerous is the registering of crofts which is merely a stepping stone to further legislation. The loan element of the Croft House Grant Scheme (CHGS) should be reinstated.
3. Occupancy Requirements. I want to see this removed. Who can guarantee six months of occupancy? Circumstances change. If the site is on poor agricultural land then it doesn't affect crofting agriculture. To suggest that the local authority should police this requirement is outrageous.
4. Crofting Regulation. I believe owner-occupiers should be treated the same as tenants. Sublets should be strengthened. I have difficulty with the Absentee Initiative. I don't think absentee tenants should be forced to assign their tenancy as long as they are subletting it. I do agree with giving the Crofters Commission more power to tackle the problem of underused or non use of croft land, currently controlled by the landlord. A tenancy shouldn't cease but the land should be used by the active crofter if the demand is there. Maybe I'm wrong but I can't understand the waste of grazing.

There are twenty-eight questions in this form. If you don't agree with any of the proposals, use a separate paper and explain why. Don't underestimate your own views.

September
Margaret Fay Shaw
Ten years ago, I wrote a short story about Margaret Fay Shaw, a two year old Highland heifer and today I'd like to tell you about Margaret Fay Shaw the 6th and 7th.

Margaret Fay Shaw the 6th is a spirited, lively, four-year-old black heifer. She was always the leader of her group but had never been handled other than in pens. She knows her name but when I shout Margaret, the whole bunch comes running, answering to that name. In April Ruth and I were over on Vallay Island, checking the cows and feeding them treats, when Margaret ran away towards the high dunes. Having made sure that the rest were all OK, we sneaked away to follow Margaret because I thought she might be about to calf. Her black colour was easily spotted and in five minutes a dun calf was born. I could see she was behaving normally, licking and loving it, and so we left her. Next day we went over. I went to the calf, touched it and realised that it hadn't stood up. Neither had it suckled, but as I always carry artificial colostrum powder and a thermos of warm water, I gave it a litre of that. I lifted the calf up but her legs couldn't take her weight. I put some nuts on the ground for Margaret and lifted the small calf towards her udder. At the same time I was worrying about how to get them both home because a free range cow needs to be penned before being put in a transport trailer. It was also over-confident to think that this lively young cow would allow me to get close enough to put her teat in the calf's mouth. I kept saying her name, and then to my amazement she never moved and I managed to do it. The next day, the calf sucked the teat providing I held her up. As the calf was safe and sheltered in the grassy dry dune and couldn't fall anywhere, we got into a routine. I decided to leave them on the island. Margaret would be with the other cows when we arrived, but I just had to shout Margaret and she would leave the rest and follow us to the calf. We carried the nuts in an old brown shopping bag that was very unlike a feed bag, otherwise the whole lot would follow us, and only Margaret knew what was inside. I would start to unzip the bag and she learned to open it herself. When we needed to rest the calf, we closed the zip to make her treat last longer. It was a week before the calf could take her own weight but it took exactly six weeks before she got up on her own. Young Ruairidh was with me that day when we stood at the top of the dune and looked down and shouted Margaret. I could hardly believe it – the calf struggled and got her hind legs up and then with a huge effort stood there all alone. Then along came Margaret and we weren't needed.

We kept an eye on her and three days later, when we arrived at the dune and looked down, there was no calf. She must have walked a bit and being a dun colour was camouflaged. Ruairidh had better eyesight, and started laughing. He said, 'Look over there about a hundred metres away and you will see her. I saw her from the other side, but I wanted to see your face when you realised the distance the little calf had walked.' It was like a miracle.

So what had been wrong with the calf? At first I thought she was premature, but without seeing her, Geert our Vet diagnosed white muscle disease, a lack of vitamin E, and prescribed selenium. He saw her at five weeks and doubted his first diagnosis, thinking it was more like brain damage. She is now in Kyles, can run and do everything like the other calves but her neck is squint so she does look rather odd. There's no way I can catch her.

Now I ask myself questions – was it wise to spend so much time and energy and fuel on a calf? Should we have put her down the first week? I think I did it for Margaret who showed me that her fear of being touched by human hands was less than the powerful instinctive knowledge that the calf needed human help.

October
Machair Project
Crofters who are fortunate enough to have machair ground will be pleased that the Machair Project has the go-ahead. It has already been in the press. This won't be a competitive scheme like RSS but more like a top up support. It will start in January 2010, but there will be public meetings at the beginning of November. I would expect those with an interest to attend.
Included in the budget is money for goose control. As you all know they are increasing every year, and unless the numbers are

reduced, this Machair Project funding will be wasted. I hope to be at the Goose National Review Group in Edinburgh in November. The Machair Project will be high on the agenda. Any of you good with a camera could help by taking photos of damaged crops and standing geese. Send them to me to show The Goose Group. Please make the effort to give me more evidence.

Blue Tongue
There have been no cases this year in the UK which is very good news. In November, a decision will be taken on future vaccination which I hope won't be necessary. It has been quite a nightmare; stressful on man and beast, a lot of work and cost, including waste. If compulsory vaccination continues, we must get ourselves organised properly for a voluntary scheme.

Crofting Reform Bill
The Crofting Reform Bill responses are being seriously discussed in government. One thing is sure, the Occupancy Requirement got thrown out the door. In a few weeks we will see how well the Minister has listened to us.

Electronic Tagging
It seems the electronic tagging of sheep will happen soon unless the EU makes a U-turn. Norman Leask, SCF Parliamentary spokesman, fought hard on our behalf and being a Shetlander he understands the uselessness of it.

Don't count your Chickens
I wasn't planning any more chickens this year, but the broody hens were cleverer than me. On the same day, one appeared with eight chickens and in the afternoon another with ten chickens. It's like working in a children's nursery, making sure that they're all safe and warm.

December

The Scottish Crofting Foundation is now called The Scottish Crofting Federation. I hope you will all agree that the word federation is a better reflection of this organisation.

Bull Scheme

Sarah Allen, who is in charge of consultation, will report to the environment minister next week. My instinct tells me that she will be very supportive. Cattle producers have strongly defended the bull scheme and I believe that Roseanna Cunningham, the minister, realises what a critically important issue this is.

Machair

Only machairs within designated sites qualify for this scheme which starts in Jan 2010. It seems that Paible is not one of them. You would think one machair is just as important as the other and that all are worth conserving.

Geese

There was a good turnout at the meetings. RSPB and SNH were left in no doubt that the geese are destroying traditional crofting and unless there's a huge reduction in numbers, this four year project will be a waste of money. In November, I attended the National Geese Management Review Group (NRMRG) meeting in Edinburgh. As if that wasn't enough, there will now be a paid policy review on NRMRG to be finished in August 2010. They are getting paid to research, interview and report while crofters are watching their livelihoods being destroyed. It's past my understanding.

It's been a very wet autumn, but we had weather like this forty years ago too. I remember moving the stooks[1] four times before eventually getting them dry enough to be put into *toitean*[2]. Also taking the *toitean* home and having to shake the snow off the top sheaves as late as the first week in November. Then we took advantage of two days of good weather and worked through the night by moonlight. Of course there were more hands available then.

Margaret Fay Shaw, the calf, is thriving. She still has a squint neck but, although small, is in good condition and looks happy and content. The chickens have all survived and I found a good home for both broods. Despite being hatched so late in the year, they seem to be coping with the rough weather. Probably because nature provides them with more feathers.

How did we manage without mobiles? Although my mobile can do anything but make a cup of tea, I don't know how to use the technology. I make and receive calls and that's my limit. When I phone Angus, it is usually important, concerning cattle. Anyway today I phoned and all I got was, 'Please leave a message.' When I met up with him, I complained, ' It's no use to me to leave you a message if a cattle beast needs attention.' Angus Og came up with a solution, 'I suppose he could set his phone like this: press one if there's a cow in a bog, press two if there's a bullock on the road, press three if your jeep won't start and press four if everything is okay.' Maybe someone could run a course for people like me.

It was high time for crofting to be accepted as a subject and the young crofters at Liniclate School are doing well. I have read their handbook The Crofting Year. It is very well written. It might be a good idea to sell copies to the tourists who would like to know what crofting is all about.

We are approaching the end of 2009. I hope it will be a joyous time and may God give strength to those who will find this season sad and difficult.

2010

February

Alexander and the barnacle geese

When I was a child, my brother and sisters and I played a game called 'Bang Bang'. We were cowboys and indians, and our guns were the jaw of a sheep. We would climb on the walls of the thatched roof barns and byre, and chase each other. If you were seen and the other person shouted, 'Bang Bang!' you fell and pretended to be shot.

Games haven't changed! Alexander has a beautiful strong plastic toy gun and harmless plastic bullets. Need I say who bought him that? It's usually grandparents that spoil children, but Alexander's big brother, Fraser, is the obvious one. Alexander isn't interested in cowboys but he is in shooting geese! I wonder if he talks about geese in school. When I'm with him, he points to flocks of geese on the ground or in the sky.

'Granny, did you see them, are you hearing them?'. As if I needed reminding. 'Granny, see these in Angie's field, they're barnacles, there's a thousand there and Fraser says I mustn't shoot them. They're special, Granny, that's what Fraser says.'

'Yes, Alexander.'

I had to turn around so that he couldn't read my mind because if that gun were real and Granny knew how to use it...I'll leave the rest to your imagination!

Newsletter

The Agricultural College newsletter contains important and interesting articles including the one about Lyme's disease spread by ticks. If the moorland was grazed properly, ticks wouldn't thrive. Our cattle graze on moorland all summer and must get immune to ticks because we never get problems with them. Muirburning[1] would also help to control them.

Responding to the article on calving difficulties, I can't understand why year after year crofters continue using large continental bulls. There are plenty of native breeds and their beef is tastier.

BSE changes

In July 2011, the BSE rules are changing again. Cattle can be slaughtered up to 72 months without having to test the brain stem. This will save time and money in the abattoirs. Currently, every beast over thirty months must have the spinal cord removed. I hope they have extended this age. Having to remove the spine, which holds the spinal cord, spoils the carcass. It's done to sheep over twelve months too, so you can't get a double loin chop from a ewe or a wedder, nor can you cook a sheep's head.

Goose schemes

There is news just in about the goose schemes and, as usual, we have a fight on our hands. Reading between the lines, I think once the Machair Project is over, the goose management funding will stop. I will do my best to get correct information for our next meeting.

March

Alexander's hens are laying well with last year's eight pullets all laying brown eggs. They are a bit small, but tasty.

When I killed the last of the young cockerels, Alexander was concerned I had killed the mature Rhode Island Red.

'Granny, you mustn't kill the chick cockerel,' as he calls it, 'or we won't get any eggs.'

Poor Alexander finds it all a bit confusing. The pullets are beautiful, each one completely red like the cockerel. Belle, the pet hen, is as fit as ever and still the boss. She demands her oatmeal in her own private dining place.

Alexander had not been well for two weeks – the usual bugs that spread around. He ate nothing at all for a few days so it was good to hear him giving me an order for pancakes. I was so pleased that he was feeling hungry that I fired the Rayburn a bit much and the pancakes were a bit overdone. He ate five, and he didn't complain about the near burnt bits.

His mother and I had been wondering what could tempt him. I used to make him semolina pudding, but not recently.

'Alexander, would you like some semolina?' I asked.

'What's that, Granny?'

'Remember, Alexander, the white pudding I used to make.'

'Oh yes, Granny, and you used to put Auntie Jessie's bramble jam on it.'

So he got his semolina, but he had to make do with blackcurrant as the bramble was finished.

The Shooting Group is doing well but resources are limited, and the meagre budget will not be sufficient. Fuel is expensive, and ammunition takes a big slice out of the budget so we are looking to the Government to come up with more money. Now that the SNH and RSPB admit that we have to reduce numbers of geese, the funding is tighter than ever. Why do they not just get on with it? They don't need to worry about goose extinction, but they SHOULD worry about livestock extinction with the geese destroying most of the grazing and polluting the animals' feeding grounds.

And the barnacle goose is a major problem too. In this small township of Kyles, there are about two thousand every day. I feel sorry for my neighbour who has lambs coming soon because his field is second from the road, very green and damp, and so a paradise for geese. What will his sheep eat during April and May?

The barnacles come in October and go away at the end of April. The Scottish Crofting Federation is starting an online petition soon but I am going to have a paper one, so look out for that. Maybe then the Government will take notice.

April

A Dry Winter

We've experienced an exceptionally quiet, dry winter. There's been very little snow but the hardest frost for many years. And, because it was dry before the frost, there was very little water in the machair hollows so breaking the ice didn't help. Our homemade troughs burst and unless the cattle could reach the river, the only way to access water was by digging holes in the damp areas. The frost that we had in the late 70s was worse. I remember Angus and I going to the river and trying to break the ice with a pick-axe and we failed.

Margaret Fay Shaw

Our calf Margaret Fay Shaw has survived the winter but still walks with her head to one side. Her mother didn't get near a bull until September, so Margaret has lots of milk. One day, during the coldest spell, I went down the machair and found her lying helpless in a hollow. Obviously one of the bully cows had given her a nasty blow and she was thrown and slipped on the ice. Her hooves were higher than her body but she just couldn't get up. I helped her but she had been there all night and staggered and fell back down again. Thanks to the mobile phone, I got young Ruairidh to help me and of course he managed to hold her up and in half an hour she was back to normal and followed the others to their bale of silage. But first she had a belly full of milk from her mother.

I wondered why there was a glass dish on the doorstep next door. I brought it in and Michelle said, 'That's Belle's oatmeal dish.' I thought she was talking about a new pet but it turned out that Belle was the hen that was wounded by a roll of wire. Alexander, aged four, brought her to his house and named her. When I feed the hens, Belle and her companion leave the others and follow me to the sheltered side of the house where they know they will find oatmeal in a scallop shell. If I leave the door open, she'll try to come in, or if the dog is lying in the doorway, she will just walk over her. Even hens will answer to a name. They're not so brainless after all!

Scottish Crofting Federation membership

I'm still disappointed at the reluctance of crofters to join the SCF. Those who are in ESA[2] schemes and then automatically got into RSS[3] would not have qualified if we in the SCF had not fought hard and lobbied on their behalf.

Goose Management Policy

We had our goose meeting in March. From last November, the government is paying for CTC consultancy, along with the British Trust of Ornithology, to review National Goose Management Policy in Scotland. Sue Evans, an independent consultant, met with the local goose committee, and willingly extended her visit so that the SCF could hold a public meeting. She came to Kyles on Wednesday and I took her round the areas populated by barnacles and greylags. She took photos that conveyed the urgent need for a solution. Frustration is turning to anger and this year we will only accept action, not words.

July

Alexander and the eight chickens

I'm not sure if I'm dreaming, but I'm actually sitting on a cushioned garden chair in glorious sunshine in my sister Agnes's garden in Ayr and I'm doing nothing! Tomorrow I will go to Renfrew to my sister, Flora, and will spend Thursday and Friday at the Royal Highland Show. I haven't been for five years so I am quite excited. On Thursday the Highland cattle are being judged and on the Friday I'll be proudly admiring the Hebridean sheep brought all the way from Hougharry by Marie Clare Ferguson.

A dry winter

Despite the driest winter on record and constant east winds, the corn crops and grass are growing fast. The vegetable gardens are looking well, and calves and lambs have had a good start. For as long as I can remember, there was always a pool of water in every hollow on the machair, but last winter the hollows were dry yet the ground is not parched and the wild flowers are blooming, bees are buzzing, and the corncrakes are crek-creking.

Annag Ormiclate

It's always difficult to wean the calves off the cows that stay on the machair all winter. The fields are too close and the calves will just jump the fence and run back to mummy. For the first time, seventeen-year-old Annag Ormiclate is not in calf. A few Saturdays ago, her bullock calf was taken off her and with the others was put in the cattle float and transported to the other side of the bay. She cried and paraded all evening and I felt so sorry for her. Sunday morning, she was near my house and still crying. Then there was silence and I thought that at last she was accepting that she had done enough mothering. I opened the door to let the dog out and let the cat in and there in front of the house was Annag and her big calf. He must have jumped the fence and swam across the sea when he heard his mother's cry. He was sucking her lovely milk and she was licking him. Well, what could we do? We couldn't upset her again. Next day, Annag, her calf and the other geriatrics went to their summer grazings. In the autumn, she will wean the bullock calf and neither will be hurt.

Alexander's eggs

While I'm away, my grandchildren are looking after the hens. Alexander loves them and collects the eggs. I didn't want any chickens this year and as I grabbed a broody hen to put her under a box, Alexander came to investigate.

'Granny, when are we getting chickens?' he asked.

Before I could tell him my intentions, his charming smile got the better of me.

'How many chickens do you want, Alexander?'

'Eight, Granny.'

Eight is his favourite number – the letters in his name and the number on the football shirt that Fraser bought him. So I chose eight big eggs and he carefully put them one by one under the clucky hen. I told him that the hen would need to keep them warm by sitting on them for twenty-one days.

'That's twenty-one sleeps,' he said.

'Then the egg will crack and the chicken will come out.'

'That's a long time, Granny.' He stood still and then jumped up. 'Granny, can't we crack them now and let them out?'

Hopefully the 30th June will see the arrival of Alexander's chickens. Deciding not to have chickens is as difficult as keeping New Year resolutions.

August

Alexander's twenty-one sleeps were over and the eight eggs were ready to crack open. He and his pal, Archie, threw away the football and came into the old hut where the broody hen was sitting in a fish box. Next minute, four little balls of fluff appeared from under the hen and the boys' faces lit up with excitement. One black chick and three yellow ones, too delicate for the boys to hold. We left them there for the night and the next day I took away the four unhatched eggs. Three were rotten and one had a dead black chick inside. The rest had to be moved to a vermin-proof, wooden small shed. Alexander and I put the four chicks in the kitchen organic bin and I carried the hen and we put them safely in their new home. Alexander put water in a shallow dish with small stones to prevent the chicks drowning. He sprinkled oatmeal on the wooden floor and proudly admired his own chickens.

Next day, one of the chickens was dead, probably trampled by the hen, but Alexander's disappointment didn't last long because that afternoon when we took the eggs in, we found another hen with nine chickens. This was unexpected, and after all the waiting and disappointment, Alexander ended up with twelve chickens instead of eight.

Margaret Fay Shaw

Margaret Fay Shaw, the squint-necked yearly calf, has to take second place now as her mother gave birth to a beautiful black heifer.

Canna Anna

Canna Anna is a young, dun coloured cow, her mother a black cow that came from Canna. Canna Anna calved in July 2009 and her male calf was left with her in Vallay all winter. Gille Dubh, the black bull, was taken away in October but in November, I saw Canna Anna looking eagerly for him, and I assumed she was not in calf.

When all the Vallay cows were moved from west to east and were gathered for lice treatment, Canna Anna was down as 'empty, with old calf.' In June, all the cows came off Vallay and the older calf came to Kyles because he didn't need his mother any more. There were a few late calves born on Griminish which my friend, Ruth, checked early every morning. In early July, I told her that there were no more calves to look for, but on the 11th July, she phoned me at 7am.

'Is there a dun calf at Airigh mhic Ruairidh that we haven't tagged?'

What was this all about? I thought hard. The only two cows that hadn't calved were empty.

Ruth phoned back. 'The calf is running around Canna Anna.'

'Well,' I said, 'she's definitely empty.'

'Canna Anna is licking it and it's trying to suck her.'

Again I said, 'He won't get much there.'

Ruth convinced me that Canna Anna was licking the calf and there was white froth around its mouth.

I was amazed and surprised and couldn't believe my own judgement any more. I had to go over that afternoon and make sure that neither of us was dreaming. I took a tag over, but the calf was so fast that I needed Fraser's long legs to catch it. It was three or four days old and Canna Anna had been hiding it in the long heather. Now nothing will surprise me!

September
Machair Scheme
The goose scarers are working hard. I hope oats, rye and barley can be secured for next year's seed, and that more stacks can be made.

Stock Clubs
Sheep and cattle sales are fast approaching and from what I hear, prices should be good. Still, the stock numbers are decreasing and maybe we should get our heads together and revive stock clubs. Moorland is going to waste and the underused crofts could play their part.

Alexander Update
Alexander started school last week and is in the Paible Gaelic Medium. I hope his teacher is a patient listener. He has many interests, from dinosaurs and tractors to cows, fish, football and of course his chickens!

He is down to eleven chickens. It was my fault. I was busy checking cows and the weather changed, and I hadn't put up the pallets to shelter their pen. The heavy rain was too much for one. When Alexander came home from school, he dug a hole and buried the chicken.

Margaret Fay Shaw
I wished I had my camera with me last week. Margaret Fay Shaw, the young black cow, was fondly licking her squint-necked 2009 daughter and her 2010 daughter was sucking her creamy milk. A lovely picture.

October
I welcome money being available for crofter training. Maybe we older generations could draw up a list of skills that shouldn't disappear including how to pluck a goose and get it ready for cooking. Actually with all the geese getting shot just now, a

student could be employed in a full time job! Cleaning sheep's offal and making black & white puddings could be another. Not many crofters under forty know how to hand-bind a sheaf of corn, lift stooks, make and stack corn ricks, and judge the timing of each job. The same applies to hay making and we could squeeze in how to milk a cow!

Two years ago, an eighteen-year old agricultural student spent six weeks on our croft. He came from Luxemburg where his family bred Highland cattle and managed a campsite. He was a fantastic help. He could handle machinery, do fencing, joinery jobs, build chicken pens and clean the chimney. He knew how to handle cattle and did it well. He helped weed the vegetable patch and could even cook. But something he never dreamt of doing in his six week practical training was helping to dig a grave in Kilmuir Graveyard. This job he undertook with great respect and wrote it up in his report for his college tutor.

This August, Serge and his girlfriend came for a two week holiday. I took them as far as Eriskay and she couldn't get over how beautiful the islands were. They remarked often that the people here were friendly. I don't know why Serge called it a holiday because he worked just as hard as he had two years ago, but they were happy. But I was sad to see how they were treated at Benbecula airport security. The girl's English was poor and she was upset. Her hand baggage was x-rayed twice and she was taken into a small room to have her cosmetics tested. She even had to remove her scarf and I'd been admiring how she wore it. I saw no TLC as I watched from the doorway. They phoned me from Luxemburg and told me they weren't stopped in Glasgow or London. I felt embarrassed given how much they admired and loved this peaceful place.

It's harvest time. My grandson, Fraser, broke a croft record this year by cutting grass silage in July in a field that has been covered in seaweed every year since before he was born. Then again, a second cut of forty-two bales in September. Of course this was an exception and the corncrakes had plenty of nearby fields to spread their wings.

Alexander's chickens are now free to roam without me worrying about seagulls taking them. Eleven are very healthy and we can start to find out if they're male or female. Alexander is happy at school, but he loves Saturdays and Sundays.

November
Alexander and the black puddings
Alexander loves black pudding. I hadn't made it for a couple of years. We slaughtered wedders in Lochmaddy so I went out and got the necessary ingredients. It's much easier to remove intestines hygienically when doing a home kill, but in the slaughterhouse, the vet has to cut off the narrow ilium and fluid escapes and stains the rest. It has something to do with BSE but what does it have to do with sheep? I eventually brought everything home in large containers. It's a good job I didn't have to stop and give anyone a lift! As I drove down to the beach, Alexander was playing with his toy tractor so he came with me.

'Granny, what have you got there? Is the sheep dead?'

It took some explaining and his curiosity grew. We dropped one lot in the sea and while I emptied one intestine, he held on to the rest in case it floated away.

'Granny that's a lot of poo. What is it?'

So I tried to explain how the grass was converted into all of this. As we let the unneeded pieces float away, Alexander was amused watching the seagulls diving down and having a treat. Why on earth do we bin all this and watch the seagulls go hungry? It's something to do with the word *pollution*. Alexander was a great help and I let him cut some of the skins and empty the poos and rinse them thoroughly, and then put them in a clean basin. Unfortunately he was at school when I made the marags[2] but when the big rumen[3] was cooked, he had just come home and again he was full of questions.

He looked at the smaller puddings and said, 'Granny, that's the kind mammy gets in the shop. I'll come and have dinner with you and will you make me some?'

So the slices went in the frying pan and Alexander sat at the table.

'Granny, have you got tomato sauce. I need that with my black pudding.'

I apologised that I never ate it with sauce.

'Will you cut the plastic off?'

I reminded him that they weren't plastic but the skins he had helped me wash in the sea. I think the whole operation was a bit much for his five-year-old mind but when he'd finished eating, he gave the empty skins to the dog so then he understood it wasn't plastic. It's a lot of work but I was delighted and appreciative when Joan Dunskellar offered to help me, and being young and fit, she was a real bonus. So between Alexander and Joan, the two-and-a-half days of work were pleasant and interesting. Mind you, I was quite shattered and to crown it all, my new collie puppy, Rocky, arrived. Not an easy task to deal with!

Last night I was going to relax and write something for the Am Paipear but the phone rang. It was the young voice that no granny could resist.

'Granny, will you plug the goose Fraser shot and will you cook it for me?' He mixes 'pluck' with 'plug'. 'I'll help you, Granny.'

It was pouring outside so I was rather hesitant.

'Granny, put your oilskins on.'

There was no getting away with it. He did help, and then watched me singe it with the small gas primus, and remove the guts. He was quite amazed to learn that birds don't do wee wees. The giblet fascinated him and the grit inside it added to his curiosity.

'We can eat that and the liver and the heart because that's meat.'

When I cut the head off, he remarked, 'it's definitely dead now'. He thought that when the head was still on, it was still a 'teeny bit alive'. So tonight he has invited himself to eat with me and on the menu is roast goose, gravy, home grown carrots, turnips and potatoes and a mug of milk. I do hope that there aren't many Uist vegetarians reading this. I wouldn't like the Am Paipear to lose customers!

In September, Andy MacLellan cut some of our corn in Vallay with his binder. The weather was perfect and it was great to handle the golden sheaves of oats. Thanks to Angie MacDougall, the geese didn't do any damage, and by the end of September we made seven ricks, all bone dry. A week afterwards, we took it to Kyles where we made a small stack. It was good for the young lads to see how work used to be done. Alexander tried hard to throw sheaves on the trailer but it was too high and they all fell back down. I hope more will do this next year.

It's getting close to bonfire night and I wish there was a ban on fireworks. Despite people being aware of upsetting animals, dogs especially, they still continue. Animal welfare is always high on the agenda, and yet this ritual is never ignored. The young ones are energetic, gathering wood for the bonfire. Pallets are burned that would provide shelter for livestock, and then they'll go to the shop to buy kindlers!

My eldest sister, Flora, once told me how she and our mother used to drag every bit of driftwood off the beach and saw it with a bow-saw. They were short of peats with the men away at war. I always imagine them sawing the wood, and hate to see this waste.

December
Winter has arrived but despite the beauty of the snow-covered land, I hope it will soon go. The children are having fun but when I tried to make a snowman for Alexander, it just wouldn't stick together.

SCF in Uist and AGM
In January the Scottish Crofting Federation executive, Patrick Krause, and one of the directors are coming to Uist to explain

the new Crofting Reform Bill and the community mapping of crofts. It would be a good opportunity for us to hold our area AGM. You must also decide if you want to amalgamate the three South Uist branches. We must have some new willing office bearers, especially young crofters, for the future of crofting is in their hands.

Goose meeting
Recently I attended the local Goose Committee meeting, now led by the Machair Project which is overseen by RSPB. The Chairman, Bill Dundas, is the Ministry of Agriculture Area Manager and based in Oban. At last I feel we're getting somewhere. There was definitely a strong sense of urgency about future goose control plans. And we decided to write to the government about the barnacles.

Red Cross
At this time of year, we all get letters from needy charities. Listening to the radio last week, an ex prisoner-of-war talked about his awful experiences and said he would have been dead if it hadn't been for the Red Cross. I was told that during the war, Dr Julia Macleod, the late Dr John's mother, milked Jersey cows, sold the milk and donated the money to the Red Cross. Dr Julia was very generous and enterprising and certainly made a huge effort helping the Armed Forces. When I was at school I used to go round the houses collecting for the Red Cross. It's a pity that it stopped.

Alexander's chickens
Alexander's chickens have had a rough time. Fraser's new puppy killed two of them and Alexander was in tears. At first the puppy never bothered the chicks so I don't know why she suddenly attacked them. Alexander wouldn't do anything until he and Fraser had buried the chickens.

It must have been really wild on Thursday night because when I went out in the morning, the chicken hut was on its side and

I could only see one chicken. With a strong north wind, the only place they could have blown was out to sea. I felt miserable. How could I tell Alexander that five had disappeared overnight? I looked everywhere, and with a heavy heart I fed and watered the hens and the brood of three in the other wee hut and locked them all in for the day. I have another cosy hut where Piper, the old dog, used to sleep and now the hens have nests there. I went in with a bucket and what a lovely sight – there huddled together were the five missing chickens. I never told Alexander because he's had enough upsets. I just hope the remaining nine will survive the winter.

Soft-hearted bulls

Most people think that bulls are ruthless animals, obsessed only with mating. Gilleasbuig is a two-year-old dun bull that we bought last February. A black 2009 bullock joined him in February and spent the summer with him and twenty-two cows.

A few weeks ago, it was time for him to retire for the winter so he and his group were penned. He was put in the box and transported to the bottom of the Committee Road. Gille Dubh, another winter retired bull, was there with three bullocks. As soon as Gilleasbuig came out of the box and realised he wasn't very welcome, he jumped the fence. Angus put him back but he jumped again, but his mind was made up and he was heading back home. What an instinct they have! He had made it quite clear that he wasn't going anywhere without his mate. So back in the loading pen went Gilleasbuig and with him the black bullock.

This time, Angus took them to Griminish and he went off happily, but made sure his wee pal was following. I think Gilleasbuig's pal will enjoy more years than bullocks would usually get!!

This year is almost gone and when I look back I realise how fortunate we are to live on the islands. I wish you all the Seasons Greetings and the best for 2011.

2011

February

Alexander goes Strictly. And wants to challenge Goliath

Despite us getting the best weather, it has been a white, slippery and dangerous few weeks for traffic and for crofters with livestock. It was wonderful to see it all thaw, although I'm sure the children won't agree. During the heavy snow, as I was passing Paible School, I saw a very happy scene. It was play time and all the children were almost finished building seven huge 5ft tall snowmen close to each other, looking as though they were having a chat! I've never seen such tall snowmen and the children were laughing with joy.

Highland Cattle Society International Gathering 29th Sept – 2 Oct

We met in the mornings at the Mitchell Library where we listened to interesting speakers from all over the world. One hundred visitors and thirteen countries were represented, the majority from EU countries but also New Zealand, Australia and America. We visited folds near Glasgow and on Saturday attended the annual Pollock Park Show. It amazes me how passionate those people are about Highland Cattle when here in Uist they're rejected. Mind you, some of the best continental suckler cows are descended from the Highland breed. One of the highlights of the gathering was a Gala Dinner in Stirling Castle where there was a fantastic atmosphere. On the menu, one of the starters was Uist hot smoked salmon. Was I proud of that! However, I chose the soup!

The Countryfile Uist programme was well received, the scenery was magnificent, and a good advert for tourism. Alexander had a great time at the dance, not intimidated by Matt Baker who was second in the last series of Strictly Come Dancing. He was up on the floor with one of his school pals, blonde at that, and maybe the young Cornish girl, Isolde, found Alexander more charming than Matt Baker. Neil MacPherson may be getting a couple of dance students in the near future!!

Alexander went to church with me one Sunday evening. This was his first introduction to a normal service. I was a bit apprehensive because he is not a child who likes to sit still and quiet for long. He joined me singing psalms and listened attentively to the Minister. The sermon was on David and Goliath and it was just what an adventurous wee boy liked to hear. I didn't even have to give him a sweetie!! When he got home, he told the whole story to his mother. He remembered Goliath was over 9 ft tall so decided he was as tall as Daddy and Fraser together. Since then he has pestered Fraser to make him a sling but Fraser wisely knows that a sling and a five year old wouldn't be a good mix.

Two weeks ago, we had an interesting visit from a young Estonian couple who wanted to purchase old bloodlines of Highland Cattle to join their fold. The weather was foul but it didn't deter their enthusiasm. Fraser had been preparing bulls for Oban in February but after getting an offer no one could refuse, three bulls and fifteen heifers went to a quarantine farm to prepare for the long journey to Estonia at the end of the month. They also purchased five black heifers from Margaret and Angus MacDonald. They left here in United Auctions' trailers. We hope they will breed well in Eastern Europe.

April

Spring has really come and isn't it wonderful to have calm, dry, warm days and a slight breeze from the west? It gives strength to man and beast. I've seen some lambs and although it's early, maybe they're getting the best of the weather and a good start. The grass is growing and if it wasn't for the geese, the ewes would have a good bite.

Talking of birds, I saw the claw prints of a sea eagle where it had walked down a sand dune. The print was the same size as my size 6 wellington! It would be interesting to watch a close encounter between a massive 8ft wingspan sea eagle and the 5ft span of a Highland cow's horns. I don't think that would ever happen but I mustn't tempt fate! I've seen the sea eagle a few times standing in the marran grass and in flight.

The hens are laying really well and Alexander is quite excited collecting the eggs from his five pullets. He likes to know which eggs are his, and which are the older hens' eggs. Sometimes it's not easy to convince Alexander, especially when I don't know myself! The cockerels were enjoyed by all the family and Alexander accepted their fate as normal and wasn't upset. I think it's better to tell them the truth. I got a beautiful Rhode Island Red cockerel from James, the postman in Eochar. Father Roddy watch out because Alexander will challenge you at the Show!

Actually, this year I'm looking forward to chickens with a 'sire' like that producing Rhode Islands. They make good table birds, so already my sister, Jessie, is looking forward to a meal of chicken cooked in butter. However, we mustn't count our chickens before they've hatched. I must put rings on Alexander's pullets so that I will always recognise them.

My sister in Renfrew was watching Countryfile with Shona, the terrier, as usual lying in her bed. When they showed the Oban Sale with Adam Henson talking to me, Shona got up wagging her wee tail and ran to the back door. She thought I was coming in. She was all excited, and when I didn't enter, she ran back and stared at the TV. Last time I saw her was in September. Isn't it amazing how she recognised my voice. Rocky, my puppy, is getting more obedient and Fraser's Bee Bee stopped chasing the hens.

May

In the April Am Paipear I read Lachlan MacInnes' article and honestly it was so affectionately written that, had he walked in my door, I was probably blushing! In the 50s most girls assumed that boys forgot about you the next day, but it is lovely to know that Lachie didn't. I do apologise to 'her indoors' when I say that it is nice to think he still remembers and, dare I say, still keeps a wee corner in his heart for me.....thanks, Lachie!!

Alexander

After church last Sunday, Alexander came to see me, so I grabbed the opportunity to tell him about the sermon I had listened to. It was the Easter message, and usually his attention span is short-lived but he was surprisingly interested. He was very sad that Jesus was killed on the cross and asked so many questions. He pondered over the big stone being removed, his only vision of removing huge stones being by machinery, so that was a difficult one to understand. At five, a child has a fantastic imagination. However, he accepted all I told him and was happy that Jesus rose again and still lives. He was sitting at the kitchen table and suddenly he looked towards the draining board above the sink.

'Granny, what are those five eggs doing there?'

That was the sign that he had heard enough for the time being, but I know he won't forget. It was quite funny to watch his wee face, listening to all I was saying and then changing the subject so perfectly.

His big brother, Fraser, broke his leg a few weeks ago, so now, after school, Alexander is at my door. .

'Granny, will you come and play football with me?'

So we play until he scores ten goals and then we stop. Maybe with all this practice, the North Uist team will want me as a reserve!

Mairearad the Yellow Highland Cow

If anyone wanted to know how a Highland cow's head should look, Mairearad was the perfect model. Her horns were fine and they curved where they should. She really was beautiful. However, she had one weakness, her tummy was over-stretched so that she always looked as if she was pregnant with twins. She was thirteen years and heavy with calf and we always checked on her. Ruth and I saw her perfectly happy, walking among the rest one day, but the following day when Angus went to check the Vallay cows, Mairearad was lying down and unable to get up. As she was in a hollow, it took four men to get her into the trailer and she was taken to a small field near home. She managed to move herself along the ground but couldn't get up. The vet was baffled but suggested maybe the cause was a back injury. She ate well and showed no signs of sickness, and I could see the calf moving inside her. She stopped pulling herself around and changing position probably because the calf was getting heavier. Ruairidh and Angus Og and Fraser would turn her regularly. In spite of this care, she started getting a pressure sore on her brisket. I spent a few days then, wondering whether the calf could be saved with poor Mairearad in such a bad way, and considered a caesarian, or perhaps induce the birth but that would have taken quite a few days. Then, as I was taking her a drink of water, I was amazed and so pleased to see she had actually started to calve. In her down position, we guessed there would be a problem, but the vet was too busy to come. It wasn't difficult to decide who was the most skilled 'midwife crofter' in our area...and a few minutes after my phone call, along came Angie Maclean. In a matter of seconds, Angie had a rope on both the front legs, and after a gentle pull, discovered that the head wasn't coming so he had to turn the head so that it would be delivered with the feet. Slowly, he let the front legs slip back, then, holding another thin rope, he managed to get it round the head which he got into the right position. Then, with a steady pull, the calf arrived, safe and sound. It was great to see how professional Angie was, no panic, he knew what had to be done and he did it. I was the only one panicking! The last time we had to help with a calving was twenty years ago when a cross Jersey-Highlander presented with the calf hind legs first, and it was Angie who delivered that one too.

Mairearad the cow couldn't help giving birth, but she gave the calf a good welcome, licked it dry, and was happy to have it beside her. We hoped she would get up once free of the calf's weight, but sadly she didn't and her sores weren't getting any better. We didn't want her to suffer, and although she was fond of the calf, he couldn't reach her udder and feed. Angus had to put her down. Luckily she already had a beautiful daughter who has calved, another female. Mairearad's calf is black and male, so choosing a name was easy – it had to be Angie! As Sarah's horses don't need the luxurious stable recently built near my house, Angie has the benefit of this 5 star accommodation. But I know wee Angie would prefer to have his mother and to be outside with all the other calves. I was saying to Angie (Maclean) that he should have been a vet, but if he was, we wouldn't have him close by! Sarah and Alexander adore the calf, and my niece Jessica, who was on holiday here at the time, was in her element feeding him. I don't know if Angie (the calf!) would like to move to Renfrew but Jessica could always hire him out as a lawnmower. At least he wouldn't break down or run out of fuel!

June
Update on Croft Characters
Alexander's 2010 chickens are laying every day. Belle the hen is still around and comes to the door every day to get her oatmeal treat. All the hens get oatmeal every day but Belle has to eat her oatmeal in a scallop shell. Many of the hens are brooding but she is laying her white egg every day. It would be nice to see her with chickens. I'd been watching a hen appearing every morning before the rest were allowed to get out, then one morning I saw her sitting beside the hen house and I could see one chicken beside her. I ran out thinking this was a normal brood, but I discovered there was just one and it was very cold. I took it in and put it inside a woollen sock and warmed it up in the Rayburn. I then went back and found the hen sitting on top of old railway sleepers. Under them was the secret nest but unfortunately all the eggs were rotten. It's strange that only one chicken survived. The hen and chicken are in the small wooden chicken hut and the chicken is growing feathers. Alexander loves to watch it and

can't make up his mind if it's a girl or a boy. It's going to look like its mum with feathers of different shades of brown.

Margaret Fay Shaw, the squint necked calf, is now two years old. Despite her deformity, she's in good condition, though small. We can't risk leaving her with her mother and they must part. Mother, Margaret F Shaw, has already calved a red female, so until now the whole family has been together. Next week the two-year-old and the one-year-old will stay together, along with the other yearlings. So now Margaret has three daughters, dun, black and red. She herself is black.

Rocky is getting very obedient but I can't trust him yet among cattle unless he has a lead on. As I live beside the beach, he has a wonderful exercise yard and my old dog, Flora, and him have a great time chasing birds and using tangle for a tug of war. He always comes back when I blow the whistle. Fraser's Bee Bee is less boisterous and if she finds a hens egg in a hidden nest in the shed, she will bring it to Fraser unbroken.

I have been exceptionally busy. Problems with calves are a huge demand on time and energy. It was nice to have my sister, Jessie, and David in their Bayhead cottage for a couple of weeks so I just called in for my dinner. Jessie did get her treat of roast cockerel in butter. I had it frozen since last winter.

I'm afraid I haven't had time to play football with Alexander. He's desperate to teach me some tactics. I don't think I should try a slide tackle after Fraser broke his leg doing just that!

July
A trip to Sweden & Alexander the rebel
Can you imagine mingling among a crowd of three hundred people, all representing their country, in fact thirty-one countries, Africa, the Americas, Russia, Arctic, Asia, Oceanic and Europe? The

majority wear their national dress or at least part of it, the colours absolutely brilliant. The Terra Madre (Mother Earth) Indigenous Conference was organised by Slow Food, mainly sponsored by the Christensen Fund, hosted in Jokkmokk, Sweden by the Sami people. Myself, Morag MacKenzie and Iain MacKinnon were gratefully funded by Vodafone Foundation. Terra Madre is an organisation fighting for the rights of native people and as each country spoke of their experiences and problems, there was a dynamic passion and sincerity in their voices. If we crofters were recognised as Indigenous we would have more power in decisions about our land, we could protect our culture, and have much more control. Some may argue that many crofters are incomers. Alasdair MacIntosh, Human Ecologist writes, 'A person belongs inasmuch as they are willing to cherish and be cherished by a place and its people and being grateful unto the culture.'

Many of you will have watched the programme about the Sami people. They are unique and have their own language, a language once oppressed just as Gaelic was. Different though we were, we had issues in common. An Indian told how they were supported to improve their native cattle breed, but when the drought meant they couldn't survive without ample water, they had to return to their own breed. Maybe it's a similar story to the import of continental cattle and the decline in Blackface Sheep on Crofts!

We flew to Stockholm, then to Lulea and travelled by bus to Jokkmokk. We stopped to take photos at the Arctic Circle and saw a number of reindeer crossing the road before running into the forest. Some excitement! We were entertained in the evenings and what a ceilidh! The Sami people's singing was wonderful. The food was – well, rather different! Reindeer meat was served, but I found it very strong, quite different to our venison. There was a starter which looked like black pudding, but it was reindeer blood mixed with oatmeal and salt, served thinly sliced and cold.

The Sami took us to the Sami Museum where outside wood fires burned and thin sliced meat was frying in huge Woks. I thought it was reindeer but it was even more exotic – bear meat.

They passed it round with lovely bread. I didn't take any, but then I thought, I might as well taste it. Hesitantly I took a bit the size of a postage stamp and found it quite beefy, but I couldn't face eating meat from a bear. They don't kill many – but this was a special occasion. I thought I had a strong stomach. Maybe the Sami wouldn't like to eat our Highland bullocks!

Alexander has been refusing to go to Sunday School.

'Granny, I don't like it, it's too long, it's an hour.'

'Alexander, how long is the school day?' I asked,

'Six hours,' he replied.

'Do you like school?' He avoided an answer.

'Granny, if I don't go to school they'll put Mammy in jail.'

So I replied,' Alexander, if you don't go to Sunday School maybe the Minister will put Granny in jail.'

'No, Granny, he wouldn't do that.'

I would welcome advice from other grannies!

August

In the early nineties, I went to the Dalmally Cattle Market to buy two black Highland cows that had come from the island of Canna. I met Neil Ferguson from Barra and discovered that he was a brother of the late Donald Ferguson, and we started chatting. While we talked, twelve black yearling bullocks from Canna entered the ring. Neil was admiring them.

'I'd love to get a female Highlander. I wish they were heifers.'

I joked, 'You never know, Neil, maybe the Canna stocksman has made a mistake and one of those is female.'

That was when Donald Morrison, the auctioneer, shouted, 'Stop a minute, take that one out, it's a heifer'.

It was like a miracle. Neil was thrilled. The black heifer was auctioned and Neil became the owner of a pedigree Highland female. It's not easy to differentiate between male and female because of the long hair, especially when they're very young and you are not handling them.

A few years afterwards, Donald Morrison brought Neil's heifer to North Uist to mate with our Highland bull. I made enquiries because the calf had been wrongly sexed so she had to be registered and given a name. In those days, the Canna cows were given a name that started with that year's registration letter, like cars. In 1992 it was D so she was named Dileas by Donald.

Dileas never went back to Barra and eventually belonged to Ardbhan. She produced many calves which were sold all over the country, but her last calf, a red bull, born in 2010, is in Spain. We exported fifteen heifers and two bulls in early summer and they all had to be red. This bull calf was the best of the red ones and not connected to any of the heifers.

Dileas was nineteen years old and we were all very proud of her. When she was carrying this bull calf, we almost lost her in a bog. Luckily we found her alive and she survived the ordeal. Sadly in May this year she was put down as she had become arthritic and was beginning to suffer.

It would be nice to go to Spain and see how the bull will develop. If it wasn't for Donald Morrison's keen eye, we would never have spotted Dileas and she would probably have been slaughtered long ago. For non-Gaelic speakers, Dileas means Faithful, and indeed she was that.

Alexander's sling

Alexander and his family visited his maternal grandparents in France. He was hardly home when he appeared with a sling. At last he had his wish. He'd been begging for one since he'd listened to the story of David and Goliath.

'I'm going to kill geese,' he tells me.

This village is full of children on holiday so Alexander has lots of company. He's having great fun with Archie MacKenzie and with help from Fraser, they've built a den from old pallets and wood.

'Granny, come and see my den.'

So I followed them over and was told, 'I'll be looking through those holes and I'll see the geese flying to Kirkibost and I'll get them with my sling. Granny, go inside and you'll see it better.'

I crawled inside and stood on an old pallet that creaked with my weight.

'Granny, you're too heavy. It's time you went on a diet.'

I looked at the two cheeky smiling faces as I stumbled over the loose floor boards and laughed at their frank remarks. Little did they know that a few weeks ago, before I went to Sweden, I discovered I could still get into the kilt I wore in the early 80s and here was my charming (so I thought) almost six-year-old grandson telling me I should go on a diet. To crown it all, he took Flora (here on holiday) to see his den, but told her, 'Girls are not allowed in.'

By the way, it wasn't his granddad nor Fraser who bought him the sling, it was his dad.

September
Belle of the Ball

'Any pet that can be caged, excluding dogs and lambs.'

That's what the North Uist Show asked for and I thought of Alexander's favourite, Belle the hen. Why not take her?

Show day arrived and Alexander was so excited. And what a day! There was a strong east wind and driving rain and our hen house faces east. I had borrowed a small cat box from my neighbour so Alexander and I put newspaper in the bottom. We took food out for all the hens and opened the small hatch. The poor hens were almost blown away. There wouldn't be a problem catching Belle, but Belle wasn't there.

'Granny, where is she?'

He was frantic and so was I. I looked round the sheltered areas shouting Belle but no Belle came. I opened the door to put the basin of food inside and out of the nest she jumped, as much as to say, 'I knew I'd get breakfast in bed in this weather.' I picked her up, struggled to shut the door and off we went to the show.

Judging time was getting near. Alexander was expecting to

parade her and as he talked to her, he noticed she had laid an egg. That added to his excitement. The judge looked at the guinea pigs but decided Belle was the winner! He beamed, 'Look what I got,' holding the red rosette and he opened the envelope. 'I got £4.' He repeated, 'FOUR pounds and Belle has an egg!' I've never seen him so excited. Mind you, Granny was just as happy. When we got home, the weather was better and Belle was happy to get out of the cat box and followed Alexander to get her daily ration of oatmeal in the scallop shell.

Bovine Viral Diarrhoea

The BVD scheme will be up and running soon but we have until November 2012 to do the first test. It would be good to get rid of this horrible disease and we test the cattle when we gather them for other reasons.

Crops in areas that got flooded in May and June are lost so I hope the crofters suffering such losses will see a bumper crop on the high ground. The hay is patchy and the yield is definitely less as a result of the thousands of barnacles grazing there. The goose scaring and shooting is ongoing so I hope the crops will be protected but as usual there is nothing happening to protect the grazing. Why can't SNH and RSPB realise that they aren't doing enough? They blame Europe – so easy to blame someone else. The estates are not blameless because if they really cared, they have the authority to bring more shooting parties. The bird watchers that come here don't have any interest in watching geese and the shooting parties prefer to shoot the lovely harmless snipes. I am disgusted when I watch that. You would think they didn't have any meat for tomorrow's dinner.

Once and for all, if the goose numbers were highly reduced, the geese would have a happy life and the livestock would have plenty to eat and it would not be polluted. When I attended regular goose meetings and they discussed scientific details of goose populations I thought it made as much sense as the cackling of the greylags.

However, I do like the scarecrows that the children have made which are a far cry from the one I once made. I dressed it with an old green gabardine coat that was past wearing even for milking the cows. We had a fantastic crop of barley and the scarecrow was to frighten any birds that came along. The stooks of six sheaves were almost touching each other. I stood the scarecrow on top of the hill on Saturday and on Sunday my mother asked Angus, 'What's your mother doing among the barley on Sunday?' Of course he teased her that I was lifting sheaves, but she had a good laugh when she realised that it was a scarecrow wearing my old green coat.

The Corn Bunting Scheme is now open with the same rates as last year. Those who are on the Machair Project area can apply to them, and the rest can register with Jamie Boyle or phone me. I'm looking forward to seeing the new reaper binder working. Kenny Walker has one already so I'm sure he will give some advice. The seed drier has arrived, so that should be really interesting.

Discussions are taking place about reviving sheep stock clubs. Those of you who have been and are involved in such clubs could maybe give sensible advice. It would be beneficial for all if stock graze the moorlands. It would take pressure off the in-bye, and keep the moor grass and heather shorter thus making it more palatable for the deer, maybe encouraging them to stay there away from re-seeds. The mutton would be tastier. Tick numbers would be reduced and if the government would finance it, there could be a job for a shepherd!

October
Alexander prepares for the storm
Hurricane winds are forecast to be arriving in the west of Scotland on Monday or Tuesday. Anyone who was in Uist 11th January 2005 is bound to feel very frightened. Alexander was very busy all day Saturday helping Fraser secure any object that might blow away, and the toy tractor and wheelbarrow were put in

the big shed. His den was tied down and anchored by huge rocks. One side of the old hen house wasn't very robust so Alexander and I started a big repair job. He brought me stones so together we rebuilt the dry stone to stop the old building collapsing.

'I think you need some wood in this corner, Granny. I'll get you some,' and he held his hands out, demonstrating the size of the wood he would need. Soon he was back with very useful planks.

'Granny, I brought you nails too, and two hammers. You hold it, Granny, and I will hammer the nails.'

I placed the small plank, and Alexander took the hammer, and took a six inch nail which he tried to hammer into the wood. I tried to keep a straight face as he tried so hard to drive the nail in.

'Granny, maybe you can do it.'

After some explaining, he went for small nails which he easily drove into the wood. 'Granny, I thought the big nails would make it stronger. I don't want anything happening to Belle.'

The best idea may be for Alexander and Granny to write Santa Claus a letter and ask him to please bring Belle a new house for Christmas, and we could leave some eggs and a sheaf of corn for his reindeer (I hope Alexander's Dad is reading this!)

We were all thankful when the winds decreased, but we know that we are not immune to extremes. Although we seldom experience them, it's good to be prepared.

Many young crofters and children haven't seen corn harvested traditionally, and I believe there is a huge lack of awareness of how harvesting was done. I was pleased to hear that the schoolchildren spent a lot of time lifting stooks (*suidheachan*) and enjoyed it. This was organised by the Machair Life project and as healthy as any PE in the gym, and very safe. Health and Safety shouldn't be necessary, unless of course there are thistles in the corn! How we hated thistles then when we were hand binding and stooking. I used to love coming home from school at this time of year to help the adults fill the horse drawn carts with sheaves of corn, all the family working together. When the cart was full and tied securely, we would sit on top of the load and hold on to the ropes for safety.

The crofting courses set up by the Scottish Crofting Federation started this week, and there is a huge uptake. Seemingly the students enjoyed the first evening. It will run for ten weeks at Lionacleit school on Wednesday evenings.

November

Water, water, everywhere
Nor any drop to drink

I believe that these words from the Rhyme of the Ancient Mariner will remain in the minds of all those who were taught in the 50s in Bayhead School by Mr D.J. Boyd. Who would think that clean drinking water would ever be precious in North Uist? 'You never miss the water til the well runs dry.' I seriously think we should all reopen and clean the old wells in case a similar situation occurs.

I was in school when P.L.J.Heron Ltd started the waterworks, and as they dug the ditches nearby, we used to watch progress. Most of it was dug with pick and shovel and the squad worked exceptionally hard, often in atrocious weather. I remember when the one and only digger arrived and lots of people came to see it. I doubt if a flying saucer would have attracted more attention! The digger driver came from Skye but the rest were all Irish. It brought a lot of work to Uist and a great supplement to the croft income. The firm's men needed lodgings and stayed with families all over the west side. At one stage, five of them stayed with us in Kyles. Can you imagine catering for outdoor hungry men when the houses were without running water and electricity? En suite bedrooms were unheard of, of course. My sister, Jessie, was home and I can remember her making packed lunches every night. It was never-ending. I usually dodged and followed my father out with the horses!

My young cousin kept hearing everyone talking about the Irishmen and he had to come and see them. He stood in the doorway and was shocked and disappointed and almost in tears to discover that the Irishmen were ' the same as us,' as he himself put it! When the job was complete, everyone was delighted to get *uisge na beinne* hill water into their houses. The well water in this area was only used for cooking and drinking because it was very hard, and however much soap was added, you couldn't get any suds. It was full of lime and blocked the kettle spouts. We had to catch soft rain water for washing clothes.

In the summer, we spent days at the peats and we would light a fire and boil a kettle of hill water. It was brown with peat but that didn't bother us, and the tea was delicious. No thermos flasks in those days.

I don't know when chlorine was added to the supply and I didn't know until a month ago that aluminium sulphate was also added. Surely powerful filters could keep it clean, but then I am not a scientist. I can't understand why the lovely *uisge na beinne* has been tampered with so much that we are scared to drink it. Are our immune systems so weak that we can't tolerate a few particles of peat? Of course, the experts give us the usual jargon – bacteria, E.Coli, EU regulations, health and safety. In the 50s and 60s, as far as I know, there was only one man employed by the Water Board to do the regular maintenance, Donald W Mackinnon. If he was alive now, he would be shocked to see the tankers taking water from Benbecula and Lochmaddy.

However, the North Uist water crisis is over, and we are now back to normal, but I have some difficulty knowing what normal is supposed to mean in 2011.

I think this summer and autumn have been the wettest anyone can remember, and we've had very little sunshine. Harvesting has been a struggle as the corn was late in growing and in ripening, and the combines had to catch the small windows of dry weather. Traditional harvesting with the binder had the advantage of usually getting it in dry. If the stacks were well made, the cereal would dry as long as damp sheaves were placed on the outside row and on the very top. The Machair binder did some for us in Valley and it cut very well but the sheaves were tied too slack. However, this was the first time it has been used in North Uist so maybe next year it can be adjusted. We made 230 *suidheachan* (stooks of six sheaves).

Despite Angie MacDougall scaring the geese, every time the tide permitted, the greylags ripped the seed off 54 stooks, and that is a lot of seed. So one thing contradicts the other. It is high time the right person went to Brussels or maybe someone from Brussels can come here. I don't think they care, as long as everything looks good on paper and they have ticked the right boxes. I know RSPB do good work but it is time they realise that chasing geese all over the place is very cruel.

The mild, damp weather has been ideal for rats breeding so in addition to the geese causing losses, the rats have chewed their way into the cereal silage bales. Despite his young age, Alexander is very conscious of the weather, after all he is always hearing it being discussed. After five days of bad weather, him and I were at the chickens, enjoying one beautiful morning, birds singing, the pet calves, Angie and Hughie, sunning themselves, so I reminded him that God gave us this lovely day.

I said, 'Isn't He good to us?

'But Granny, it was a terrible day yesterday,' was his reply!

My last cat died at Easter and there was, unusually, no cat on the croft. But I was told Donald John Knockintorran's cat has kittens, so Alexander and I went down to see them a few times. However, they weren't easily caught and they took a couple of weeks to arrive here. Alexander found waiting for them difficult. 'Granny, they are taking such a long time'.

They are here now, and Alexander knows they are here to catch mice and young rats and not to be house pets.

The five chickens are almost as big as their mother and he is happy to see they have all survived. Recently he was away in

France with his mother, visiting his grandparents, and the day before he left, a hen appeared with eleven baby chicks. I can't say I welcomed them, especially in October!

'Aren't they tiny, Granny!' was Alexander's comment. He had forgotten.

December
Another year's crofting

Alexander has been busy after school helping Fraser take seaweed off the beach and shooting geese. He loves his bicycle and cycles along the road, exercising the dogs.

'Are you coming with us, Granny?'

He thinks I should run along beside him or maybe borrow Sarah's bicycle. Mind you I'm often tempted, but having broken my hip four years ago, I'm scared of falling. I haven't told Alexander that I used to cycle as a hobby when I was a teenager living in Glasgow. I often cycled to Oban and Fort William and stayed in the youth hostels. I was in the Glasgow Touring Club and went in groups of eight or twelve.

Steven Morrison caught my interest when he told me that he and his pals were going to cycle from Glasgow to Uist to raise funds for suicide awareness. It was absolutely wonderful what these lads achieved. I believe they raised about £15,000. I know Steven well and when I chatted to him during a visit to his grandparents, Ronnie and Katie Wilson, he was quite surprised that I had often cycled a hundred miles a day. I did it for pleasure but those lads had an admirable purpose and they achieved it.

During my teenage years, I also went to the riding stables at Kelvinside. I worked hard early weekend mornings with fork and shovel and then in the afternoon got free rides. In those days it cost four shillings an hour for a lesson and I only earned £4 per week so couldn't afford them. I always loved the horses and worked with them on the croft along with my father. Naturally, I get great pleasure seeing Sarah taking pride in her

83

horses. She spends hours with them and I was thrilled when she competed at the Stornoway Horse Show in August and won four Firsts and three Thirds and was awarded Third overall. I watch her riding on the beach and imagine myself in her place. I never got anywhere near Sarah's standard but there's no harm in make believe!

Last time I wrote, one of Alexander's hens had appeared with eleven chickens. We kept them for two weeks and then found a good home for them with Mairi Claire Ferguson and Calum in Hougharry. I told Alexander the chickens would be happier there as we couldn't let them out at this time of year and that young Calum would look after them in a big vermin-proof shed as he loves poultry too.

Angus Alick Macdonald.

Donald Archie Macdonald, Hugh
Macdonald, Angus Archie MacIsaac
and Donald Allan MacIntosh.

Kyles Paible
in 1986.

1954: Ewen and Sally
fetching peats.

Ena, aged 16, binding
sheafs of corn.

1956: Ena, Jessie, Agnes, Flora.

Angus's record field
of barley.

1995: Robert Quarm's rows
of Angus's peat.

Cattle cooling off in Ardbhan.

Angus MacDougall and Donald MacIsaac.

1956: Ena with her parents.

Ena's father scything.

1958: Ena with
friends on Sally.

1952: Ena
harrowing with
the two Sallys.

Angus Alick and Angus.

Angus and his
grandfather on the
David Brown.

Angus at Ardbhan.

2012

February

There's a new arrival on the croft and Alexander wants Granny to take it for a run! It doesn't have feet or feathers but is fearless and fit on moor and machair.

Alexander ran in the door, 'Granny get the food for Angie and Hughie and I'll teach you how to drive the Buggy!

And that was my first introduction to the Polaris, the all-terrain vehicle, sort of between a quad bike and a jeep. Well, I was so unprepared. His Dad shouted, 'Alexander will show you how to drive it,' and left us to it. It reminded me of when I was fourteen and my late brother, Ewen, told me to sit on the first tractor, a David Brown 25D. He put it in first gear and left me sitting on it, fortunately on the beach, and when I couldn't stop it, he shouted, 'Put your foot on the clutch and put it in neutral.' I didn't even know which pedal WAS the clutch!

Well, I guess it is good to learn the hard way. Buggy has a left-hand drive and having gone in the wrong door, I just stayed there.

'Alexander, can you drive it?' I asked.

'Of course, Granny,' he said.

He had the key in his small hand and was very methodical. 'The key goes in there and now this lever goes forward.' and with a few more instructions, away we went down towards the beach, along the dune track and, despite having to drive very close to the fence, my chauffeur was very capable and careful.

Rocky, my dog, didn't follow us because he didn't recognise the vehicle. Alexander and I fed the calves and went back into the Buggy. Alexander, feeling very important, drove me back home. I still haven't driven it myself, but hopefully next Saturday I'll take a test drive. Surely, if a six-year-old can manage, I should be able to. It made me think back to when Angus was four, he would harrow with the old David Brown and steer it perfectly through the gateways. My father would put it in first gear and hand plant the corn seed while Angus would do a perfect job of harrowing, avoiding jack-knifing the tractor and harrows. I hope Health and Safety are not reading this!

Although Alexander goes to the Gaelic Medium (GME) class at school, he refuses to speak to me in Gaelic. However, last week he came running in and spoke fluently, saying that his pal, Cador, and him were coming to Sunday School with me. It was the best Christmas present I ever got. I cannot understand why Gaelic-speaking children prefer English, especially when both parents are bilingual.

Since last I wrote, the weather has been foul with rain and wind leaving the ground saturated and water lying in hollows where I have never seen it before. Despite the poor summer, most crofters secured plenty silage but they have had to start feeding earlier to keep the livestock in good condition, and it's a long time until May. The cattle sales have been very buoyant which is encouraging but the cost of feeding stuff is high so profit is lost. Let us hope that the government will see sense and not change what hauliers have to pay under the present RET. If they penalise the hauliers, many businesses will suffer. The SCF supports the hauliers, and the fight continues. Why change something that has been so successful? The wrong decision will mean more losers than winners. It is like robbing Peter to pay Paul.

The water saga at Bayhead continues. The whole episode has left me totally confused. However we have an opportunity soon to meet Scottish Water Officials, so maybe they will have some answers. Since I discovered aluminium sulphate is added, I haven't enjoyed a drink of tap water nor a decent cup of tea.

The SCF Crofting Course will commence later in the year and I hope it will be as popular as the last term. I was at the final night and it was a sheer delight to see the students' enthusiasm. The co-ordinator, Susy Macaulay, did a wonderful job of encouraging them to attend. Due to atrocious weather, some evenings had to be rescheduled, and even the venue had to be changed once. However, all ended well, and maybe there will be more hands-on courses next term. So, all you active crofters out there, be willing to share your skills.

March

This month I must apologise to those of you who enjoy reading the 'Alexander series'. I've had no contact with Alexander for over three weeks as I haven't been well. I have shingles and I have been told that I can give chicken-pox to anyone who hasn't had it. And Alexander hasn't. As well as the discomfort of a horrible rash, you feel extremely tired.

So I thought you might enjoy a Highland Cow story. Recently, we sold cattle in Oban and one of the calves called Susan had quite a history so this story was displayed on her pen. She was sold to an Estonian and in the meantime is in quarantine for a month.

Susan of Ardbhan

It was October 1990 and we had been back from the Oban Highland cattle sale for about two weeks when the phone rang. It was Donald Morrison, United Auctions auctioneer, in his usual fluent Gaelic:

'I've got a lovely but wild Highland cow here from Canna and I think you'd be interested in her. She was booked for the sale but she didn't quite make it to the mart in time'.

'Oh yes, Donald, I do remember, a Suisaidh Dhubh, born 1982 in calf to Laochan of Shenavallie, entered in the catalogue, but I assumed she had been withdrawn.'

In those days, cattle travelled on ferries in pens on the car deck, not in trailers, and when they disembarked in Oban, they just walked along the quay and Corsons' stockmen would drive them towards the old mart.

Donald went on: 'When the Canna cows walked off the boat, this cow broke away from the rest and galloped through the Oban streets and wandered through fields. No-one could stop her. After about ten days of roaming, one of the stockmen, Ewen MacLeod from South Uist, with much patience and skill, got her penned.'

I thought, 'Poor Suisaidh, obviously the only thing on her mind was getting back to Canna because she was alone and didn't know the two year old heifers she had travelled with.'

It was an instant decision and in a few minutes Donald and Angus had arranged transport for Suisaidh to come to Ardbhan. The ferry from Oban to Lochboisdale would take seven hours. The following evening, Angus took the Toyota pick-up and the cattle trailer, which was past its use by date, the fifty miles to Lochboisdale. I thought it best not to ask how he would get her into the trailer. I supposed some locals would be around! When he was driving over the causeway that connects Benbecula to North Uist, there was a thud and the trailer came off the tow bar. Luckily there was a strong rope holding fast and somehow he got the trailer back on. I hate to think where the trailer and Suisaidh would have landed If the rope hadn't been there. Fortunately Angus was young and strong.

Despite Suisaidh's adventures, she soon settled down at Ardbhan, and in Spring 1991 she calved a lovely black heifer which we called Suisaidh. She is the second oldest cow at Ardbhan, has a lovely nature, and the strength and energy of her mother. As does her 2000-born daughter sired by Neil of Auchnacaig. We have Suisaidh of Ardbhan's 2008 black bull sired by Gille Dubh 2nd of Killochries and we used him last season. It will be interesting to see if his daughters inherit the beautiful black fine horns of his maternal side, and will his sons have the excellent conformation which his sire usually breeds. We can only wait.

On the radio this morning, there was a discussion of the Lochboisdale to Mallaig ferry. Donald Morrison spoke the truth. There is a lot of stock movement over the year, over and above the official sales. How can crates, pulled by tractors, travel from Mallaig to pens in Oban? And high-sided lorries would have to go via Lochmaddy. It would be a huge loss to crofters if the Oban connection stopped. We need an extra ferry to cover Lochboisdale to Mallaig.

April

Alexander and Rocky

Now that the weather has improved, the days are longer and I am feeling better, I am spending more time training my young collie, Rocky. I took Alexander with me one day after school because he is so keen on watching Rocky improve.

'Come on, Granny, I had better drive the buggy for you. Have you got the whistle?'

Off we go, and with two blasts of the whistle, Rocky is alongside us, and soon we are racing along the beach, then onto the dune. The three of us stop at the machair gate where the seventeen pet calves including Angie and Hughie are out-wintered. Naturally the two pet calves are always at the gate to greet us, maybe still hoping that a bottle of milk will appear. Alexander takes the whistle, and with a 'Come by' command, Rocky brings all the calves together, and then 'Away' and the drive starts. If Rocky goes too fast, Alexander will blow the whistle, and Rocky is by his side. We take the calves about 500 yards to the bottom of the field, and Rocky makes sure they go through the gateway. Sometimes Angie and Hughie break away, thinking, 'We will go to the bottom of the field when WE want to go and no dog is going to force us!' But they don't realise that the pup that used to lick their noses when they sucked their bottles of milk is now grown up and only interested in herding them.

'Granny, look at Angie and Hughie. They don't want to go through the gate. Look, Granny, Rocky has seen them and is turning them back and they have joined the rest.'

He is so excited that Rocky does it without a word of command. However, we must teach Rocky to get ahead of a fast runaway group of cattle, so Alexander and I still have a lot of work to do.

Belle, the hen, is still around. It amazes me that when I open the hatch at the bottom of the henhouse in the morning, Rocky will stand close by and won't move until Belle comes out. He ignores the rest and instead of letting her eat with them, he brings her to the house to her own private dining-room to enjoy a scallop shell of oatmeal. We have twelve Scots Grey hens, identical to Belle, so how does Rocky recognise her? The colour doesn't help him so is it the scent or her movements? Can anyone answer that?

Reflections

I attended a memorial service for the late Bill Quarm. Having never been at anything similar, I was a bit apprehensive, but as I walked into Carinish Hall that all changed. How appropriate were the words from 1 Corinthians Chapter 13, 'though I speak with the tongues of men and of angels and have not love'. There was an almost tangible interaction between the family and all who came to support them. His sons, David and Robert, spoke proudly of their father, and praised their mother for their happy childhood.

Bill was quite special to me. He was the factor and I was the manager when Bayhead shop was first opened in 1969. With the two estate joiners, Bill worked many hours, helping to change a school into a shop. He was my boss but we each did the jobs we were capable of doing.

As David spoke and showed slides of his parents' wedding day, it was hard to believe that forty two years ago, I was in this very hall, helping with the catering at the reception. They made a very handsome couple and their children have inherited their good looks.

David, while still at school, helped us a lot on the croft. When Angus built the first shed in 1986, David laboured alongside him using shuttering from the demolished wooden walls of the knitwear factory on the Kyles road.

May
Changing Times

Why do we put the clocks forward? If time and tide are governed by the sun and moon, why do we tamper with what is natural?

Considering there is so much daylight, I can see no advantage. The animals have more sense. The dogs are quite happy to sleep until 8am. The hens don't want to go into their house until 9 pm. When the spring tide is low, it should be out here at Kyles at 12 noon, but of course my clock shows me it is 1 pm. When I was a child, some of the older people didn't change the clock. Our late minister, Revd James Morrison, wouldn't so in the summer we had our services at 1pm and 7pm.

Alexander has had a very happy Easter holiday. He and his mum went to the south west of France to visit parents, grandparents and Great Grandma. He was confused about Grandma looking after Great Grandma. 'Granny, why doesn't someone look after you?' I convinced him that I was still fit to look after myself, and suggested that when he was grown up, he could look after me. He replied, 'But Granny, Dad will be there to do that.' I think he prefers things as they are now with Granny playing football with him and looking after his chickens.

The Bramble families go back and forth to Kyles until October. Lucy and Nick and their daughter, Daisy, and son, Archie, were here when Alexander returned from France. Archie and Alexander are the same age and had a wonderful time. They played long and happily with plastic swords and of course football. They even went swimming. I shivered just to watch them! Another pal is Archie MacLellan and I must not mix their names! The island one is pronouced Erchie and mainland one is Archie. 'Granny, that is the proper name.' There is beautiful honesty in the mind of a child.

The chickens which hatched in December unfortunately didn't survive the winter but they lasted long enough in a chicken house inside the big shed to waste an awful lot of my time and energy for a couple of months. A lot of the hens are brooding already, so I keep putting them under fish baskets to stop them.

June

I'm disappointed in Belle, Alexander's pet hen. I was sure she had a brain superior to other hens, but…

Alexander came in. 'Granny, here are four eggs I found in the shed. You must come and see where I found them.'

I followed him and he showed me a cardboard box containing nails and thin bolts, a mixture of metal pieces. 'That's where I found them. The hen must be very stupid.'

But the egg shells must have been very hard not to have broken because Alexander came in with an egg every day.

'Silly, silly Belle, you really are stupid,' said Alex.

I thought I had better not agree. 'Actually, she's very clever, your Belle. She thought nobody would look there, and she could hide the eggs. Maybe she was planning to have chickens.'

He was quite happy to think that Belle wasn't stupid at all!

The cats are doing a grand job. There isn't a rat to be seen in the big shed. What a clearance. I never thought they would catch fully grown rats, but I saw them in the act, so in the meantime I can put the cages away. Nine female calves were in the shed all winter. The cats lay down beside them and the calves would lick them and the female cat would rub herself against them.

Every winter we keep about five or six cows in a machair field. I call it the geriatric ward. Last January, one of the two-year-old bullocks that were overwintered next to the cows jumped over the fence and stayed in with the cows. A couple of weeks ago, when I was there, I thought the bullock was very like this old cow, so I checked his tag number when I got home. No wonder the resemblance was there. It was her son. The bullock born in 2011 was there too. So even cattle can be mammy's pets and yes, they do remember!

July

The Machair Life has had an art project going for a crofting year involving twelve S1 and S2 pupils from Lionacleit School learning how art and crofting can be combined. They were shown how, and allowed to help with, traditional harvesting and taught about seaweed and its uses. They learned the skills needed for four seasons of crofting. I was invited to meet them at Taigh Chearsabhagh[2]. There was a slideshow with the children providing their own enthusiastic commentary. Even the quieter ones were contributing. And they were honest enough to admit that one of the joys was that the project took them out of the classroom!

When I was at school, some of my classmates would be off school to help with croft work. Without machinery, children and horses were much in demand. So, back at school after being absent for two days, it was, 'Please, Sir, I was needed at home.' and that was that.

At the beginning of June, a cow calved and we put her in the crush because the calf was only sucking one teat. This cow was ruptured in 2006 when she was carrying her first calf. It happened when the Department of Agriculture made us do cross checks – checking the tags in both ears. The cows have to be penned, and Highlanders and pens are not a safe combination. This horrendous job was being done in March, very close to calving. I can still see her pushed against the wall by another frightened cow. So, being out of shape since that incident, the calves have difficulty in reaching all her teats during the first week of their lives.

Alexander was watching while Fraser and I helped the calf. We decided to milk out the calf's favourite teat and I put the milk in a container.

'Granny, can I do that, can I drink it?' He touched the milk with his lips. 'Granny, can you put it in the fridge first please?'

As a child, Fraser was the opposite. He drank gallons of warm milk. He would have a mug in his hand while I milked, waiting for me to fill it. I used to cheat by putting a small jug at the side of the Raeburn so that he enjoyed it as if straight from the cow! That evening, Alexander had a large mug of the cold milk with his dinner, and the same for two more meals.

'Granny, I love it! Will you bring a cow home and I will milk her every day.'

He was really excited about it, so with the school holidays starting soon, I am going to organise getting milk for Alexander. A few cows will allow me to milk them, but getting there before the calf is the problem. I am sure Alexander can persuade Fraser and Dad to help out! No wonder he enjoyed the milk. Shop milk is like water compared to raw milk. People my age probably wish they could still enjoy the cream on top of the milk, and have their own butter, cheese and crowdie. My mother told us that the last jug of milk in the udder contains more cream, and that was kept in a small jug to add to the cup of tea. I can't remember anyone drinking black tea in those days!

We had one out-of-school treat to celebrate the coronation. Every child and all the teachers marched from school up the Knockaline Road, then a gravel track, to Craig Hastin. Ahead of us was John Angus MacDonald playing the bagpipes. We arrived at Craig Hastin to find almost every able-bodied person in Paible there, and a picnic had been prepared. Peat fires were heating up large kettles and everything to eat was homemade. We had games, long jump, high jump, running and tug-of-war. I remember I had a new dress of my very own and red sandals, and we all had hankies and flags in red, white and blue.

August
Alexander sees red over lobsters
Alexander has a few lobster creels on the West Side. Ewan and Calum Iain find it amusing to come across creel buoys marked Alexander. None of the fishermen have a boat of that name,

'I thought all lobsters were red, Alexander,' said Auntie Flora, teasing him.

'Oh, Auntie Flora!' and his face showed what he thought of her. He shook his head and sighed. 'They will be red when you cook them'. He was too polite, or scared, to tell her she was stupid.

I'm not very brave when it comes to cooking them and I was glad that my nephew, David, dealt with it. He also took Alexander fishing for brown trout but despite his experience, he caught none whilst Alexander got two and had to get home then so that we could admire his big catch. I offered to gut them, but Mum had to see them first.

The young cat, Cissy, has four kittens. They were two weeks old before I discovered them hidden inside a huge tank where Angus stores the winter corn seed. I couldn't climb in because there were too many empty ton bags. Eventually, Sarah, Alexander and I managed to get the bags out, and we put in clean straw, food and wooden boxes for the cat to climb in and out of. The next day they were gone. A few days later, we saw them in the loft in the big shed. From there, she moved them to the loft on the opposite side, where we could get food to them. Two days later they disappeared again. This time I found them on the shore, hiding like rabbits among the sea defence rocks. I was worried they would get stuck and I dreaded the possibility of a high tide drowning them all. The mother cat had enough sense to move them, and returned them to the safety of the accessible loft in the big shed. The cat climbed up and down the ladder to eat, along with the kittens. Eventually she moved them downstairs, to the shed floor, and Alexander and Sarah have them quite tame. We have homes waiting for two of them, but if anyone wants to take the other two, please get in touch.

September
Breakthrough in Goose Control
The National Goose Management Review Group (NGMRG) is a government group consisting of representatives from various

organisations such as SNH, NFU[1], SCF and landowners. For ten years I attended numerous meetings, mainly in Edinburgh, representing the Scottish Crofting Federation (SCF). I was the only one who understood and was affected by goose damage. The rest were in paid jobs and kept reminding me that the group's remit was to protect crops, not to reduce geese numbers, and that it was an EU directive. I reminded them that while protecting crops was vital, the geese flew to the grazing ground, polluted it and ate all they found there. There were times when I could scream. At the first meeting I attended, the chairman thought he was smart, telling me that us crofters only had ourselves to blame. 'There's an open season for shooting geese from the first of September, so why don't you shoot them?' He was embarrassed when I told him that crofters don't have shooting rights. Those belong to the estate!

The local goose committee has been doing a great job, but they have to answer to the NGMRG. Two and a half years ago, the government sent consultants to review the NGMRG and we in the SCF put in a response. We gave them a true picture of what was happening, especially in Uist. Due to the recession, the Government was only willing to fund the goose schemes for two more years. The government review proved that despite all the money spent on the NGMRG for protecting crops, the goose population was increasing year by year and they knew that when government money was withdrawn, not only would agriculture suffer, but also rare species of birds would surely decrease further. In Islay, thousands of pounds have been paid to farmers for compensation to allow the geese to multiply. Here in Uist, we didn't go down that line. Now, due to pressure from crofters and farmers, the NGMRG decided to set up a pilot scheme in the four areas that are hardest hit by goose damage. In Uist, the Machair Project put in a bid to the EU and was accepted. This pilot scheme has a welcome change in its policy called Adaptive Management. From October to December geese can be shot not just on cultivated land but anywhere they are seen. It won't be plain sailing, but it's a great breakthrough.

If the so-called environmental experts had listened to the crofters twenty years ago, when the warning signs were already red, by now the very rare corn buntings, corncrakes and yellow bumble bees would be in their thousands and attracting visitors from all over the world.

Alexander has had a very happy summer holiday and has made the most of having big sister, Ellie, at home all the time. He went swimming, cycling, fishing and even sledging – not on snow but on the slopes of the sand dunes! He goes with Fraser and Sarah on the 'goose run' and as usual brings Granny the dead goose to pluck.

'I'll help you, Granny,' so we stood where the wind would blow the feathers out to sea. He pulled quite a few off the breast, but the wings were harder.

'Granny, I think I better let you do it.'

However, he did stay with me and his plucking will improve.

October
Alexander gives Granny a standing order for black pudding

Mairi is a dark brown 14 year-old Highland cow, good-natured and obedient, but she never attracted any special attention. Last spring she calved a black male, and a few days later Fraser was over in Vallay, and thought the calf was a bit disorientated and came to the conclusion that it was blind.

He and Ruaraidh came back and took the cattle-box over, brought cow and calf home, and put them in a pen in the shed. The calf had white dots in his eyes. Geert, our vet, said it was cataracts. In the shed, he was safe and never far from his mother but when he was two weeks old, he started dragging his hind legs, and was obviously in pain. This of course was joint ill, due to the lack of colostrum but a few antibiotic injections sorted that out.

Mairi and calf have stayed in the village all summer. Both are in great condition and the calf seems to see quite a lot now.

Mairi has blossomed like a shy child. She will stand and let anyone milk her, and looks at me as if to say, 'I would have been like this years ago but nobody bothered about me, just ignored me'.

When my sister, Jessie's, grandsons, William, seven, James, six, and Alister, four, heard that Alexander had been milking a cow I had no peace until I took them down the machair to see Mairi. Not many cows would tolerate three kids running around them, but Mairi seemed to enjoy it.

William was the first to have a go. I showed him how to keep thumb and finger tight so that the milk would not go back to the udder when he squeezed. He was that proud when a few squirts came, but she had very little as the calf had had his breakfast already. James was next. He was a bit more adventurous and went to both sides. Alister was more reluctant and needed encouragement, and then, when he was almost there, he decided it was, 'Yuck, yuck!' but he did grab the teat. Then Mairi decided to go and graze. The boys were happy, and the story has been told many times.

My granddaughter, Eleanor, has German students visiting her, and they had never seen a cow being hand-milked or even seen a Highland cow. Mairi was, once again, the perfect model.

I was tidying up my vegetable garden. It is satisfying to see so much ready for eating. My garden faces east, and with two months of dry east wind, I was late in planting, and the carrots took five weeks to appear. When Flora was here in July, she was wanting to do some weeding for me, but the ground was like the Sahara desert. Recently, with the ground improved, I was telling her she could come back and weed as much as she wants.

I am glad to hear that corn stacks are being made. I had looked forward to having a stack, but we lost so much from the stooks to the geese last year that it was too risky. We did make dry hay, and Angus got lovely dry round bales into the shed. I keep going in and out to look at them and smell them. Alexander's cats have discovered the warmth and comfort of them. It's good to know that no rats will be urinating in them as since I got the two cats a year ago, we haven't seen a rat or a mouse. By the way, if any of you want kittens, let me know!

Recently, I made black and white puddings. My nephew, Calum, and his sister, Morag, were here from Ayr, so I had plenty help. Calum is a chef so he was excellent at chopping the suet, and Morag, who is brilliant at sewing, did a good job at sealing the skins. Alexander came in to watch, and had forgotten eating them a few years ago.

'Granny, I don't like black pudding now.'

I said, 'OK, Alexander,' and gave him slices to take home.

Next day he came over. 'Granny, I DID eat the black pudding and I really liked it, and Granny, will you make me some every week?'

He was so excited about the *marags*. His broad smile and eagerness made the hard work worthwhile but not even Alexander's charm will not get me making them every week, or even every month for that matter!

November

Why is chicken-pox named as such? Alexander and his mum were going to spend the October holidays in France with Grandma, Grandad and Great Grandma, but it all had to be cancelled because Alexander took ill with chicken pox. He was looking forward to his holiday, and the day before he should have left, he was excited and made me laugh.

'Granny, I'll be swimming every day, but I think I am really going to miss this place.'

What a boy! You would think he was going for a year instead of a week.

Then, with the next breath, 'Granny, will you make me a scone before we go and will you put jam on it?'

Number eight has always been Alexander's favourite number. His football strip has to have eight on it and last year he had eight chickens. In August, another hen appeared and she had eight chickens. And then, two weeks ago, his little red hen, one of last year's chickens, appeared with yet again eight beside her! The young cat has had a second litter of kittens, six of them, not eight this time! Alexander thinks all this cute new life is great. While he is enjoying them, and he does help when he can, I seem to be spending hours each day feeding this menagerie of cats, kittens, dogs, hens and chickens when I could be getting on with other jobs, including the housework.

Rocky is now two years old, very clever and obedient. Now that Fraser has bought sheep, he is in his element. Mind you, he still loves to herd the hens, especially Belle.

December

Halloween night was an exciting time when we were children. The masks were home-made, usually from a well-used cotton pillow slip or a well-worn cotton vest. In those days everyone knew each other and children called on adults more often, so the game was to try and hide our identity. We didn't have bags to carry goodies in. What we got didn't reach a bag as we ate everything there and then. We got treats like dumplings, pancakes with jam, and maybe a penny or two if we were lucky. The girls often wore their brothers' clothes or if they were older they would wear their fathers' ex-army khaki trousers and always a cap. The boys would wear their sisters' or mothers' tweed skirt. There were, I think three motor cars on the west side of North Uist so a lift was unheard of. Never mind, we were well clad in winter woollens and on our feet were wellingtons or strong leather tackity boots. I was six or seven years old when I was allowed to join the rest on Halloween and I went as far as Seaview on the Knockline gravel road. There, a number of boys came down the hill and Calum Dhomhnall Eardaidh was wearing a real sheepskin mask which almost covered his whole front. I got such a fright that I screamed, thinking he was the bogey man so my sisters took me into Seaview House and Angus John had to carry me most of the way home!! I don't remember when I was brave enough to venture out again but probably the following year.

Nowadays it's a different culture. The children wear ready-made, bought masks and dress up as different characters. They don't walk much either and usually a parent will drive them around. They carry bags which get stuffed with sweets, chocolate bars, crisps and coins. When my grandchildren started guising about fourteen years ago, and I would be childminding the following week, I enjoyed helping myself to chocolate bars and jelly babies!

This Halloween, three little boys called on me. Of course I had to pretend I didn't recognise them. They were Keiran and Steven MacIsaac and Alexander. Keiran was very talkative and tried to confuse me by saying that the other boy was Archie MacLellan. Alexander was dressed as a gladiator with a helmet, uniform and plastic sword but he didn't wear a mask.

'Now I must study who this soldier is,' I said. 'I wonder if it's Alexander because he is always playing with the two of you.'

Alexander was grinning from ear to ear and his smile turned to peels of laughter.

'Do you not know me?' he asked.

By now, the other two were convinced that stupid Granny didn't recognise Alexander, and then Alexander couldn't hold back any longer.

'Granny, it's me, it's Alexander, and Granny can you give us another bag because we have to call at other houses!'

The only connection that Halloween had with crofting was the Gaelic saying, *Oidhche Shamhna canar gamhna ris na laoigh*. On Halloween night calves will be called stirks.

Thinking back, there was a lot of work done to prepare for the long winter months. Most crofters grew a lot of potatoes and the only way to store them was to build a shed made of sand sods. Every sod had to be a certain size and the spade slanted so that the sods would fit together like bricks. The structure was usually rectangular and wider at the bottom, and the sand sods had to be thick enough to protect the potatoes from frost. As most of the cattle were in-wintered, it was a similar job preparing their beds in the byre. Most of the byres had earthen floors but there was a long thick plank of wood laid below for their hind legs with a drop for their dung to avoid their hind quarters getting dirty. The sods were placed carefully, close to each other, sand side up, but the area in the stall underneath the animal's head would have the grass side up to avoid sand mixing with their corn. Straw was never used for bedding, only for feeding.

2013

February
Christmases past and present.

Christmas dinner was going to be easy with a ready-to-roast goose in the freezer and plenty beef and lamb. I was just about to take what we needed out of the freezer on Monday morning, when Alexander came running in, saying, 'Granny, you've to come and help us pluck. We've been shooting and we've got ducks and geese.' So, down went the freezer lid and on went my wellingtons and warm clothes.

'Granny, Dad and I have started plucking.' and sure enough, he had plucked most of the feathers off one duck, but the wing feathers were too tight for his little fingers. 'Granny, can you finish it for me? I've got to help Fraser feed the bullocks.' That was less boring, and warmer.

The ducks were so fat and a treat roasted. On Christmas Day we had a feast fit for a king. Michelle and the girls cooked all the trimmings, stuffing, our own vegetables and my own favourite, potatoes roasted in dripping.

Alexander had a great time as usual getting spoiled by Fraser and Ellie and Sarah. He tried to see where each duck had been shot and wanted to find the lead pellets, but the breasts hadn't a mark on them. He was impressed to find out that they had been shot in the head.

When I was a child, Christmas was very different. Before Christmas, most families would spend hours in the barn plucking hens. The feathers were kept for stuffing pillows. They were then cleaned for cooking, and well-wrapped in brown paper (we didn't have plastic wrappings then) and posted to the wider families living on the mainland. At that time, the mail would arrive three times a week in Lochmaddy, and a bus took it to Bayhead Post Office. My sisters and brother would walk to the post office on those evenings, and meet up with the other youngsters to watch the mailbags getting emptied. Most of the parcels contained apples and oranges and Christmas cakes. Dundee cake was a favourite, tea and dried fruit, and if we were lucky, some sweets. You could smell the apples before the parcels were opened. That was the only time of year we got fruit. Standing in that Post Office was so exciting. The postmaster, Allan MacLean, and his wife and the postman would call out the address on the parcel and eager arms would grab the package. We didn't have a Christmas dinner in those days, only the excitement of Santa Claus.

One Christmas Eve, my sister, Agnes, and I decided to stay awake so that we could see Santa Claus. We hung a pair of my father's woollen stockings on the brass bedposts. For what seemed like hours we struggled to stay awake, and then the silence was broken by Santa Claus coming into the bedroom and we heard him putting gifts in our stockings. We had got accustomed to the dark and as soon as he went out of our room, we jumped out of bed and chased him. We saw him run into the other bedroom which had an open fireplace, and we were positive that he disappeared up the chimney. Of course, it was big sister, Flora, wearing our mother's coat and hiding behind the door. We couldn't switch a light on as there wasn't electricity. When we went back to school after the holidays, we were so excited to tell everyone. Some of the older boys were trying to be hurtful by telling the younger ones that Santa didn't exist but we soon convinced them that we had seen him!

We didn't get newspapers but some people got them from their relations on the mainland. Our favourite was the Sunday Post. Although we didn't have Christmas dinners and Christmas trees, we didn't envy our mainland cousins. I used to gaze at the newspaper and read *Oor Wullie and the Broons*[1]. They would be sitting at a big table covered with Christmas goodies and there would be pictures of Christmas trees, holly and mistletoe, robin redbreasts and Santa Claus. Yet I accepted that we didn't have those things, that they were in faraway places. We were quite content with the magic of parcels and Santa.

You might wonder what we did get in our stockings. Usually an apple or an orange, a couple of caramel sweets. I was lucky because as the youngest, Flora would have made me a rag doll or a rabbit. Do children experience those simple joys now? Most already have more than we saw in our whole childhood.

I remember when the school canteen opened and we had our Christmas party there. Mayac MacIntosh was the brilliant cook and most of the food was produced locally. There was a beautiful spread – sponge cakes, jam tarts, Bakewell tarts and shortbread. But I didn't like rhubarb and ginger jam, and when I arrived home after school, I burst into tears when my mother asked me if I had enjoyed the party. Shortbread was all I ate because everything had rhubarb and ginger jam on it! I still don't like lumps of ginger in anything, although I love ginger nuts and gingerbread.

I am very sad thinking about the school's situation. I noticed a rhone pipe leaking badly over the back door entrance and I wished I could climb up a ladder myself and fix it. The council has to find half the costs so surely it would be less to upgrade the existing ones. Building a brand new school won't increase the roll, nor will it make better scholars. What happens if the population drops further? What will happen to the existing Paible School? Is it falling apart already?

March
Carrots and Snow
It was nine o'clock in the evening and I was thinking of having an early night. I had just taken my dogs Rocky and Flora back inside and I was about to lock my door when I heard footsteps coming fast towards my door which was then pushed open by who else but Alexander.

'Granny, Granny, have you got any carrots in the house?'

'What on earth do you want carrots for at this time of night?'

'Granny, I need a carrot for my snowman's nose.'

'Oh Alexander, calm down, we haven't seen a snowflake this winter, so how can you make a snowman?'

'Granny, I have a snowman, Donna and I have made a snowman, the lorry driver brought me snow.'

I was now getting the picture. 'Alexander, I haven't got a carrot in the house but I will get you one in the garden. You'll need to come with me as my torch isn't very good.'

So on go the wellingtons and warm jacket and into the garden we went and found a six inch long carrot.

'Thank you, Granny, you did well finding one in the dark. Will you come and see my snowman?'

Sure enough, the two of them had built a tall snowman with a hat and scarf. Alexander's Mum had been telling the lorry driver that Alexander was looking for snow every day but none had fallen. As there was so much on the mainland, he filled a few plastic bags with snow and put them on the back of the lorry bound for Lochmaddy and delivered them to Kyles. I can't remember that type of delivery ever happening before! I don't think any of the family will forget the joy in Alexander's face. Supposing he has every toy that was ever designed, none would have been as magical as the bags of snow. I'm glad he is old enough to remember that night.

The hens have done well over the winter and some carried on laying. The chickens that were hatched in July started to lay in mid January – lovely little brown eggs. There were eight chickens in that first lot, four pullets and four cockerels. When the mother hen stopped brooding in November, she left the chickens at night and went into the hen house with the other hens. In the morning, when I let them all out, she would join the eight chickens and looked after them until evening. Quite similar to a cow, when she dries up and wants to wean her calf, she will punch and kick it away but never leaves it completely.

This horse meat scandal is worrying, though it might make consumers shop at their local butchers who have struggled for years to compete with the supermarkets. So if horse meat can be sold as beef, what else might be falsely labelled? We are very fortunate in Uist because we can trust those that provide us with food. The government is trying to avoid criticism by stating that horse meat is safe to eat but admits that false labels are corrupt. There were other countries where only beef from under thirty-month cattle was allowed in the food chain and the rest was incinerated. They are still splitting sheep carcasses over twelve months although a sheep has never had BSE. The money spent on all this nonsense should have been used more wisely. I don't think anyone here would choose to eat horse meat and let's hope not many will become vegetarians either! This should be a wakeup call for governments to support local butcheries.

The Machair Life Crop Protection Scheme ends in autumn 2013. Next year we are on our own as the government won't continue funding it. However, Machair Life and RSPB are hoping to secure funds from other sources. The ongoing adaptive management scheme is funded by the Government and managed by SNH and the budget is limited to £30,000 over three years so I'm not very optimistic. Since the open season in September 2012, about 3,500 geese have been shot by North Uist estate but crofters we have to put up with 7,500 so the adaptive team has a huge challenge ahead. When the corncrakes disappear and we lose all our seed to the geese and we can't rear cattle and sheep anymore, the tourists will stop coming. Maybe we can make a living selling nettles and thistles and rhubarb. I don't think the geese will eat them.

April

The Adaptive Management Programme is ongoing and the shooting group are doing well. However, resources are limited, and there is no way the meagre budget will be anything near sufficient. Fuel is expensive, the men have to travel a lot, and ammunition takes a big slice out of the budget. We are looking to the government to come up with more money. When, at last, even SNH and RSPB admit that we have to reduce numbers, why do they not just get on with it? They do not need to worry about goose extinction, but they SHOULD worry about livestock extinction because the geese are destroying most of the grazing and are polluting the animals' feeding ground. And in this small township of Kyles, there is a flock of 2,000 barnacle geese here every day. I feel sorry for my neighbour who has lambs coming soon, because his field is second from the road, and very green and damp – a paradise for geese. What will his sheep eat during April and May?

June

'Granny, will you come and help me look after the environment?

'What! Alexander I can hardly stand on my own two feet in this north wind! And how do you expect me to do that? And anyway, should you not be at athletics on a Thursday evening?'

'Granny, I haven't got time. I've been collecting rubbish and please will you come with me in the buggy and we will take it to the bonfire place?'

Why does he always think of the rotten jobs when I am going home to get out of my oilskins and wellingtons and sit down for two minutes? When I was seven years old, I had not heard the word environment! Anyway, as usual, I gave in to him and we got into the buggy ...

'Granny, I will drive it for you?' he says, as if he was doing me a great big favour!

'OK, but watch these doors, you must only open them when you are facing into the wind.'

We filled the back with plastics and old ropes and anything that would burn, and put it on the heap. I didn't want to spoil his enthusiasm by telling him that we can't burn plastics because it is bad for the environment. Of course, I praised him for being tidy and never let on that I knew the environment was the least of his priorities, it was the chance to drive the buggy down the beach and he wanted a co-driver! He had probably been nagging big brother Fraser to accompany him and Fraser had had quite enough struggling in the gale. But Alexander saw that Granny was still outside. I wonder what the job description is for a

Granny?

The cows are in good condition, but again the dreaded white scour has reared its ugly head. I wish someone would come up with a cure because it is such a common complaint. In 2002 we lost a number of calves and our vet advised us to send a dead calf to the laboratory. It turned out to be cryptosporidium, the same bacteria that were in the Glasgow water supply from Loch Katrine a few years ago. However, we now know the reason. ValJay Island was getting fenced into three areas, and the cows had to be locked into the one where there was plenty of water, but much of it was stagnant. Then the next attack came in 2011 then again in 2012 and this year. It was the ragwort. They were eating the nuts so close to the ground that they were eating new ragwort without us or them realising it. Three were infected this year and we lost two. Number two was calved by Cashan, the tamest and most lovable cow in the whole herd. When the scour was picked up, the calf got the old-fashioned pink tablet and some Life Aid which it drank from a bottle. On the second day, it looked better, then on the third day, I went over, and sadly it was stretched out. When I got to the calf, a large gull-like bird with brown wings, maybe a skua, flew off its back and I was horrified to see the calf was still alive with a hole the size of an apple eaten away under its tail. Its eyes were untouched, but I knew he was too far gone. The scour had obviously weakened him. The only thing to do was to shoot and bury him. Another calf had died a similar fate because of a raven. Never before have I seen that happen. The mother tried to chase the raven, but it kept circling and trying to land. There are too many ravens in Uist.

I was reading about the new water treatment in Bayhead called chloramination which involves mixing ammonia with the chlorine. What will they be dosing us with next? Wouldn't it be great if they would just use a filter to stop any beetles or small lumps of peat coming through, like they use for milk in dairies, and give us fresh, cold, chemical-free water.

116

An aunt in Kyles, Morag, had worked most of her life in big houses on the mainland, cleaning and polishing, and was extremely particular. No dust or sand was allowed to settle on any surface. Everything was washed with green soap and ammonia was always added to the water. Anyone going near a shop was ordered to bring back a bottle of ammonia. Dishcloths and towels were steeped overnight. Floors were washed with ammonia. So Auntie Morag and ammonia ensured that everything was sterilised and spotless. I can't believe we are now drinking ammonia! Auntie Morag could have saved the trouble and expense of buying all those bottles if she had lived until now!

May
Crofting News
I met with Drew Sloan, Chief Agricultural Officer, who was on a flying visit to Uist trying to get crofters' views on the next common agricultural policy. There will be a meeting in Balivanich in June and as the future payments will be area and not historic, he is very interested to know how common grazings operate. His other priority is to make sure new entrants to farming and crofting will qualify for subsidy, so I think you should all make every effort to get as much information as possible before then.

Another mad press release came last week. The EU is trying to put a ban on arable farmers and crofters using their own seed for planting and making it illegal to sell to others. This would be disastrous for us when our own home grown seed is definitely the most suitable for our soil. They are also trying to stop vegetable growers from keeping their own seed. Of course this is coming from certified seed growers and it's very worrying although I fail to understand how such a rule could be policed.

Alexander has been working hard taking in the peats and helping to feed the young cattle in the shed. His pal, Archie MacKenzie, was in Kyles for a fortnight so they were up to all sorts of pranks. A German/French TV company was here for a week so Alexander was being a novice star. He and I were filmed plucking a goose and Alexander had such a job pulling the wing feathers that it looked more like a tug-of-war! The film will be shown all over Europe so might encourage more tourists next year.

August

'Maybe you would like to be a vet one day, Alexander?'

'Maybe, Granny.'

'If you did want to be a vet, you would have to stay in school for a long time, and work hard at university.'

His usual beam turned serious, and he looked down as though he felt guilty. 'I don't think I could do that, Granny.'

This conversation took place when I told him that one of his six red pullets had died. He wanted to understand why she had died, and he very much wanted to see her. So I told him I had put her in a cardboard box and that he could help me bury her. He had a good look at her, and then ran back announcing, 'Granny, I think it was dehydration. Or brain damage.'

Where did he get all that from?

Two weeks ago I found his two-year-old red hen outside, sitting on eggs in a large tub that used to grow flowers. Last year, the same hen hatched eight chickens in the same tub. I guess hens like to brood in the same place every year.

The day of the sports, six chickens appeared. When I found the hen in the first place, I was very tempted to take the eggs away, but my maternal instinct was too strong. So, although I have reared many before, I didn't want any this year because fragile chickens are prey to predators when out getting grass and sunshine. I am very happy to be giving them to the poultry-loving Ferguson family in Hougharry.

Nature is amazing. That young hen sat for twenty-one days in a container under a small bush right next to the big shed where men and machinery and tractors are coming and going all day. Dogs and cats are also around. Rocky knew she was there and just kept watching her. All that activity, and she never left those eggs except when I would see her in the yard in the very early morning, eating and drinking before anyone else was about.

Valley Island looks wonderful with its beautiful wild flowers. Tourists and locals go there every day. But I would like to meet the lazy, irresponsible, horrible person who was there last week and left a gate open. The gate is in perfect condition, the bolt goes into the post as it should and it has a loop rope and extra rope knotted for safety. There are eight cows and two calves over there, their breeding life over for various reasons like lameness, mastitis and old age. How could anyone sane do that? Surely if you find a gate shut, especially so securely shut, you close it after you. Those ten animals walked across the strand to Griminish. They would have spent every summer there, so if the gate was open in Vallay, of course they would head over to join the others. Yesterday, three of us spent over three hours getting them back across the strand. The bull at Airidh Mhic Ruairidh jumped the fence and was ready to follow them had it not been for Fraser's youth and fitness chasing him back in with the young cows. We had to be rather cruel, driving the lame ones to keep up with the fit ones who were desperate to get back to Griminish. The old cows could have been drowned.

I am sure all crofters are missing the skips. The reason they stopped making them available was supposedly to do with Health and Safety. There is a lot of fly tipping just as there used to be before the skip scheme was introduced. Nobody seems to know where to put crofting waste like broken tools, bits of fencing, old ropes and such. As for the bins, there seem to be too many of them, and too many rules and dates governing them. What a shame we lost the skips and gained the bins.

September

They say cats have nine lives and I think the same can be said about hens! About six years ago, Belle, Alexander's pet hen, survived after being chased by a dog and getting stuck for two days inside a roll of rylock² and injuring her legs. Three weeks ago she got chased and injured her legs again. Alexander and I watched her losing her balance and becoming timid and vulnerable. That day the two of us had work to do in the garden. We transplanted leeks and then tried to kill the caterpillars, but the cabbages and cauliflowers were too far-gone. Alexander collected a handful of caterpillars and like most wee boys was enjoying squeezing them. 'Maybe Belle would eat them Granny?' so since his hands were full of caterpillars, I went for Belle. However, Belle wasn't interested in a caterpillar lunch so we left her in the safety of the garden for the rest of the day.

Evening came and we had to decide what to do with Belle. Should we put Belle down? That didn't appeal to either of us. 'Granny, maybe Belle will get better, and I know where to put her where nothing can hurt her. Granny, put her in Sarah's stable.' So now Belle has four star accommodation: a bed of straw, a bowl of oatmeal and barley, a dish of water and, would you believe it, Michelle has even given her green tea which she drinks to her satisfaction. Four days ago she seemed to make a great recovery so Alexander is delighted. Maybe soon she can join the rest again.

There's a hen and six chickens in the pen every day and they go into a small wooden house at night. I move the pen to clean grass every day. On Sunday, I was ready to leave for church and luckily looked out the window. One of the chickens was outside the pen. It must have escaped through a hollow under the wooden frame. I went out dressed as I was, hat and all, and walked towards the chicken pen. Some of the little chickens have escaped before and I've caught them in a corner, quite simple. However, this time the broody mother hen went berserk when she saw me, screeching and almost trampling her other five trying to get away from me. No way could I catch the escaped chicken, so all I could do was open the pen hatch and out she came. The six little chickens followed her and they all ran into their little house where they felt safe. I was amazed at the hen's actions. She couldn't have recognised me and must have thought I was going to harm her chickens. I never thought hens were that observant. We all know that cattle get used to the jacket you wear and don't like strangers, but a hen is not expected to be so clever!

There's plenty of activity just now on the crofts with silage making in full swing. It's great to see such heavy crops and good grazing. This should be a good conditioner to help the livestock winter well. I still think that it's wrong to have strict grazing restrictions. Cattle don't like grass when it's long and rank, especially in wet areas where it loses its nutrition. Harvesting dates are ridiculous. I grew up learning that you harvest a crop when it's ready and when weather permits, not when someone sitting in an office dictates. Of course, the birds and the beetles are more important than crofters' livestock!

There are many ticks and their bites can develop into Lyme disease. I don't remember being worried about ticks when we were young. Is it caused by too many deer closer to homes, a lack of muirburn, or is there a new species of ticks? As children we used to love pulling them off each other and off the cows when they were getting milked.

November

It was a privilege to be invited to attend and take part in discussions at the Eighth Annual Forum of the European Co-ordination Let's Liberate Diversity in Basel, Switzerland.

Twenty-eight countries were represented and 178 people were present, some from as far away as the Philippines, Mali, Macedonia, Bosnia, Croatia, India, Iran, West Africa and Peru. All were small farmers or representatives of farmers, and all passionate about what they were trying to achieve. This year the title was From Planting to Plate and the main subjects were seed and livestock.

It was amazing to discover how everyone knew about crofters and warmly welcomed us. Governments are trying to dictate that small producers cannot store their own produced seed for planting the following year, not just corn seeds, but vegetables and fruit as well. They insist it must be certified and come from mainly large scale producers. The farmers want to continue selling, planting, and swapping their own seed, just like we do here with oats, rye and barley. They do not want to bring in foreign species. The Peruvians told us how they got the support of three thousand small farmers and won their fight. The livestock discussions were similar, and there were stories about small poultry farmers anxious to keep their rare breeds.

My responsibility was to explain how cattle breeding has changed on our islands to satisfy the market and to explain the problems young crofters and new crofters have obtaining land and Single Farm Payments. In addition, the lack of government support for moorland grazing.

We had one afternoon sightseeing. Switzerland is a beautiful country, so green and mountainous. We could see the snow-covered Alps in the distance. We visited an orchard where 350 different varieties of apples grow. I don't think I want to drink or even look at apple juice for a long time!

We visited a rare breeds farm where we saw some of the original Brown Swiss and the old type of Simmental which looked much smaller than what we see here today. There were four Highland cattle there too.

When the conference was over, I travelled by train to Cologne in Germany where I was met by our farmer friend, Axel. I was there in 2002 when I was on the Eorpa[3] programme talking about the Over Thirty Month Scheme. It was great to see the Highland cattle that we had exported in 2006 and the young heifers that left here early this summer. It was lovely to see Morag come running towards me when I called her in Gaidhlig[4]. I wonder if she was homesick!

In Germany, I was trying to forget about hens, chickens, cats and dogs when my mobile rang.

'Granny, it's Alexander, and I want to ask you something. Granny, you know the old wheelbarrow that is upside down, well, I found a hen hiding there, sitting on eggs, and if the chickens come, where will I put them? It is one of my red hens, Granny.'

I assured him that all his red hens were together before I left home so, as it takes twenty-one days for the chickens to hatch, she would be OK until I came back.

I am glad to say that when the twenty-one days were up, no chickens arrived and the hen abandoned the eggs. Alexander was disappointed, but chickens in October are not to be welcomed. Poor Belle, Alexander's pet hen, died last Sunday and Alexander was in tears. To be honest, I was sad too. She was special and even knew her name, but she was in her teens, and she had a good long life. Alexander and I buried her on the machair in a spot that he chose.

It was good to be home and Rocky was happy to see me. He had been very bad when I was away and fought with Ròn, Angus's labrador. Ròn bit him above the eye and Geert had to stitch him up!

It was a very memorable eight day trip, and my only disappointment was that my luggage never arrived nor did I get it back until I was in Glasgow.

December

Last week I enjoyed listening to the inter school Gaidhlig debate about the referendum. They were excellent and I think the judges made the right decision. Sgoil Lionacleit did well convincing the others that the referendum was definitely not the most important issue on the minds of young people. I quite agree, and despite its importance, we older people have other priorities too. I'm a bit fed up with it every day on the radio and television!

The crofting register workshop was held on 11th November and a few townships are now deciding to go ahead. It is better to start now, and any boundary disputes can be sorted when all the shareholders are together. Although it isn't compulsory, any crofter assigning their croft to family or anyone else, or maybe de-crofting, will be required to register the croft now. In other words, nothing can be done unless the croft is registered officially,

We hear a lot about seaweed these days. The new developments by Uist ASCO at Crogaire quarry will be a huge boost to the island's economy. It doesn't seem that long since the Johnson family planted their forest and those trees will now be producing the heat for drying the seaweed, a natural, environmentally friendly method. I hope this exciting enterprise will be a huge success.

When Sponish factory in Lochmaddy was in full swing, tons of seaweed was being cut every week and it was suitable work for crofters. In the late seventies, quite a few people in this area collected the long stalks of tangle and most went to the South Uist factory. I did a lot myself, and Angus used to join me after school. A strong north west wind would bring tons ashore at Arnol so you had to work very hard to secure as much as possible before the wind changed direction and it would all be gone the next day.

Alexander is taking seaweed off the beach with his tractor and loader, perhaps a Massey Ferguson 135. I'm not very good at recognising tractor breeds. He's made a huge heap already.

'Granny, Dad says you want a loader full of seaweed for the hens,' he said. He was happy and feeling very important now that he could drive the tractor and loader. 'Granny, the tide is going out so I'll go now and get it.' Shortly afterwards, he returned. 'Granny, I've got the seaweed, will you come out and show me where you want it?'

After some manoeuvring, he put it exactly where I wanted it.

'Granny, the hens can scratch away now and they won't need to go down on the shore when it's windy. I'll bring them more when they need it. Granny, you know the track you take when you go onto the beach and it was full of seaweed, well I've cleared it all for you in case you get stuck.'

At last, thanks to HRH Prince Charles' support, the government is recognising the importance of retaining island abattoirs. The Prince's Countryside Fund will help to finance this three year project. This is welcome news for those that are struggling to survive. A pity the Prince isn't our environment minister!

2014

March

It has been a wild and wet winter, in fact the worst in living memory. We expected wind and rain, and maybe snow, but this winter was something else. The Met Office blamed the extreme weather on the jet stream which has travelled north-east away from Britain. Watching the floods in the south of England, we can be thankful that we have a natural drainage system. It is a good job that the river through Paible was cleared in the summertime, or else the fields would have been seriously flooded. We also had exceptionally high tides, even the neap tide. The rain was constant. We have had unusual south-easterly winds. The sea has been furious.

My house is very close to the sea. If Angus had not built the sea defence after the big storm in 2005, I am sure it would have reached my door. I was thinking of Alexander as a toddler, sitting in the tele-handler with his father. The tele-handler lifted the huge rocks and placed them together like a dry stone wall. Alexander was fascinated. When people asked him what the tele-handler did, he put his two arms up like he was lifting a heavy weight and clenched his teeth, trying to copy the action of the great machine!

Calving is in full swing and soon it will be lambing time. Crofters who had their cows calving in December, January and February must have had a hard time out in such dreadful weather yet it is amazing how hardy a new-born calf is, provided it gets plenty of colostrum soon after birth. We have used bottled colostrum for years. It is very handy because there is no mixing to be done. Every time we check the cows, we take a thermos of warm water. If colostrum is needed, we just add the water, shake it well to dissolve, and screw on the teat. It is great that we can now buy these bottles locally at Simpson's. Back when I used to milk a house cow, I would freeze the colostrum to be ready for the following year.

Now that the days are stretching, Alexander is enjoying being outside after school. He is feeding Fraser's sheep and they come running to him now. He also feeds the Texel ram, whom he calls

Basher. And he takes his tractor down the beach and drags driftwood home. I think every generation has enjoyed that. We went beach-combing when we were children. We would drag the wood above the high tide mark to be collected later. Some people would put their initial on the log.

'Granny,' says Alexander, 'when I grow up, you won't have to buy any food in the shop. I'll be shooting geese and I will take Dad's boat to catch fish and lobsters and mussels. Dad doesn't like hens but I do, and I will look after them so you will have eggs and cockerels to eat.' So it looks as though I will be well provided for! Alexander thinks that while he is growing up, Granny is going to stay as she is... God willing, I hope we can share his dreams.

The Gaidhlig radio programme Air Chuairt has been nostalgic, especially for my generation. I have a faithful fan who worships my voice as though I was a pop star! This is a four-legged fan, my sister Flora's Cairn Terrier, Shona. She thought I was in their sitting room and kept running around looking for me. She comes here every July and I suppose I spoil her. Dogs are such intelligent creatures.

June

A couple of years ago, every township clerk got a letter from the councillors wanting to know our views on a Crofting Community Buy-out and as far as I know, the large majority of crofters wanted to remain as they were. I was quite surprised when recently the issue was raised again, this time titled Community Buy-out. Legally, us crofting tenants can buy our crofts at fifteen times the annual rent but few have taken the opportunity. Even as an owner occupier you are still governed by crofting law and don't have shooting or fishing rights. Laws change and sometimes owner occupiers can't access grants available to tenants. A buy-out is a good opportunity, especially when the landlord has no interest in the estate or tenants which isn't the case with the Granville family. My understanding of this movement is that it's easier to get funding from the Government and the EU or other funding bodies if the proposed development is community owned. I don't know why our estate is compared to others like some in Harris and Lewis because we are very different. This is a strong crofting community. I admit that it isn't as vibrant as it used to be, but then years ago we didn't have 9 – 5 jobs and those who stayed on the croft concentrated on crofting. But those were hard times and who would want to return to these days? First of all, North Uist Estate is not for sale so this would be a hostile buy-out. If it was agreed we go down that road and the buy-out was successful, we would still remain tenants. We would then be governed by a board or trust of ten elected people which would change every three years. I can't imagine that being something to look forward to. Would those elected be really interested in developing North Uist? Would crofting be in their blood, and would they govern North Uist any better than the present North Uist Ltd? We can't expect to have a *perfect* landlord but the Granville family live here and they create a lot of jobs. Obviously they love this island. If we have a grievance, we can approach them face to face and as far as I know they don't stifle developments. There are many developments in North Uist, undertaken by hard-working, strong-willed, determined people. Are we going to destroy a family just because it would be easier to get funding? And in a few years the funding bodies may make it harder for communities, so where would that leave us? I believe there may be a ballot and a feasibility study. Why waste money on that? At the meeting it was very obvious that the large majority were against a buy-out. I have my own personal grievances – I wish the estate would shoot more geese, deer and ravens! Let's just be thankful for what we have and work together to make North Uist everyone's perfect island.

Alexander is very work conscious now. He did a lot of rolling the ground after the seed planting and harrowing. The fields have

130

to be level and flat to make them safer for the harvest machinery. He feels important doing 'grown ups' jobs! 'Granny, Dad's tractor is stuck in the wet field, but don't worry, I'll get him out.' He had run home, picked up a towing rope, got on his own tractor and away down to the beach. In a matter of minutes, the tractor was free. He loves the sheep and lambs and looks after his two pet lambs after school. He is still very interested in the hens.

I had a laugh a few weeks ago. The four chickens that hatched last October started laying in March. They picked a stupid nesting place outside, behind a gate. Alexander always collects their eggs but I had to pull the gate back to allow him to reach them. This time, one was sitting on them.

'Alexander,' I said. 'Just put your hand underneath her.' Which he did.

'Oh Granny, she's laid an egg in my hand and it's wet and warm.'

I don't think I have ever experienced that! He handed me the four eggs and I said, 'You can't get a fresher egg than that, so we will give them to Margaret Westford. She loves a fresh boiled egg.' He had a good giggle over that!

I have had an unusual birthday treat. Fiona, the Spanish student who stayed with me a year ago, remembered my birthday. She phoned me in the afternoon to see if I would be at home at 6 pm but never mentioned my birthday and it was the last thing on my mind! At 6 pm a car pulled up and there was Fiona with four other music students attending Benbecula music college. They had an accordion, two guitars, a harp and small bagpipes. They played me Gaidhlig tunes and Fiona sang. What a beautiful voice she has. Jessie and David were here then and they joined the ceilidh. I was thrilled. We are often ready to condemn young people, but it was so thoughtful of them to take time to come here and entertain me.

Recently we put eighteen Highland bullocks in a field on the Committee Road. The grass is always greener 'on the other side' and after two days they wandered out round the fence at the loch onto the moorland common grazing. Sadly one never made it and died in a crack in the peaty ground. The rest arrived near the main road at Claddach Kirkibost but that wasn't good enough for them – they spotted the green, sweet grass at the roadside and then they were really difficult and jumped fences. Angus and Seumas Boyle got them back. Many thoughtful people phoned early and late in the day to let us know they were on the main road and that was very appreciated.

July

In May, Richard Bramble was on holiday at his house in Kyles. He is very outdoorsy, loves the windy weather and goes kite-surfing. I get nervous watching him going at high speed below my house. Alexander's two pet lambs are in a pen in front of my house. I had just fed them and usually they lie down afterwards, but when I looked out my window, there they were, standing side-by-side, heads up, staring out to sea. I then spotted Richard. So even the lambs thought he was crazy, bobbing up and down with his kite high in the sky. I watched the lambs' heads following the kite. They stared at each other, as if to say, '*Mee Beag*[1], are you seeing that? What is it?' And the other would reply, '*Mee Mor*[2,] I think it's a huge bird, but there's a big fish tied to it. I hope it doesn't attack us.' At the same moment, Rocky, who was also watching, went over to the lambs' pen to assure them that all was well. He lay down and they came as close to him as the wire would allow them. The three of them lay down and had their usual snooze. I wished it could have been filmed. Walt Disney could have made another million dollars with our lambs and Rocky.

Alexander's Rhode Island cockerel died recently. When I went in at 6 pm to collect the eggs, he was inside, lying on his back, dead. I thought a small dog or ferret had killed him, but there wasn't a mark on him. The only thing I noticed was that his crop was empty so I assumed he had taken a heart attack. He was so fat and heavy. Maybe he was over-fat. Alexander was very sad when I told him.

'Granny,' he said, 'we will have to get another cockerel. I will look around.'

'Alexander,' I said, 'we will get another one soon, but it will be difficult to find one as quiet and gentle.'

'Granny,' said Alexander, 'you will need to get another one right away or we won't have any eggs!'

'Alexander, we will have eggs alright, but we won't get any surprise chickens, and that isn't a bad thing!'

He looked at me very puzzled. I don't think he could work that one out. Anyway, no more was said. Alexander would like another Rhode Island, so if you know of one available, give me a call.

June has been a wonderful month for crofting, fishing and tourism. It has been calm, warm and sunny with enough rain to make things grow. I have never seen such good growth at this time of year. I hope that the corn doesn't ripen too early with silage cutting restricted until 1st August. If it loses its 'green' it also loses its nutritional value. The cattle and sheep are enjoying the good grazing and should be more than ready for the autumn sales.

In the late 40s and 50s all the sheep would be on the moor and the cattle would be herded. In this township, the cattle would cross to Ard Heiskeir during the spring tide, and would then be brought back into the cattle fold to be milked and locked up at night. The calves would be tethered near-by and would be fed out of a bucket. Some people might think it cruel taking the calf away at birth, but when the cow calved she didn't see the calf, it was put in a wooden pen where someone would dry it with straw. The cow then licked the jacket of that person, and bonded with them as though they were the calf. If that person had to be away, they would leave the familiar jacket with whoever was going to do the milking.

We spent most of the summer holidays at the peats. I had better not start thinking about that lovely tea, accompanied by scones and home-made butter and eggs. Flies would fall into your cup, and you would just lift them out with a spoon or a finger covered with peat. We also collected cockles and caught flounders, and *saithe*[3] and *cuddies*[4]. One of the least favourite jobs was hoeing the potatoes, but if we were told to do it, that is what we did.

The Royal Highland Show is over. It must have been the highlight of the year for the Ferguson family of Dubhairidh, Hougharry. There was very strong competition, but their Hebridean gimmer, Dubhairidh Yule, won second prize and their three-year ram Knox Quiring won second prize and followed with the coveted male Reserve Champion. They sure put North Uist in the limelight.

August
Richard Lochhead, Cabinet Secretary for Rural Affairs and the Environment, was in Uist last month, canvassing for the referendum. Some of us representing the Scottish Crofting Federation (SCF) spoke with him and raised issues. He was a good listener and took notes so we can only hope that he will act. It's good to know that the crofting grants scheme (CCAGS) budget is definitely remaining solely for crofters and will not be opened to smallholders. That would have watered the scheme down. Points were made about housing grants, young entrants, simplifying application forms, rough grazing payments, support for small black land crofts and, of course, we didn't forget the geese! I believe he understood our problems.

Soon all the crofters and farmers in Scotland will be receiving a letter with the details of the new Common Agricultural Policy (CAP). It's a pity the transition period from historic to area payments will take five years. The SCF favours area payments and soon we will know how the policy is going to be implemented. I am glad the calf premium is going to continue. It was us in the SCF that fought hard to have it in the first place because the powerful National Farmers Union (NFU) did not support it. We designed it so that the first ten calves would secure a higher premium to give crofters a good incentive to get cows back on the croft after numbers drastically decreased after the Suckler Cow Subsidy stopped. There will be a subsidy for ewe hoggs or gimmers (female sheep that have not yet produced lambs) maybe £80 each. It is wise not to pay for older ewes or we would be back to the days when sheep were kept until they dropped. This new incentive will keep flocks young and healthy.

Alexander's hens weren't long without a cockerel to protect them. Catherine Major read last month that the Rhode Island had died so she delivered an absolute stunner – a white cockerel with a black tail. He is tame and has already made friends with Rocky. Alexander's mainland pal, Archie, was here, so when the cockerel arrived both of them gave the hen house a good clean. Unfortunately, Alexander was away at show time so neither the cockerel nor the pet lambs got there.

September

Recently, my sister Jessie and her husband David celebrated their Golden Wedding Anniversary. Family and friends enjoyed a lovely evening in the Stepping Stones Restaurant where charming staff served a delicious meal. We are fortunate to have a local, family-owned restaurant of such a high standard. Alexander and his cousins, William, James and Alister left their dinner plates clean although Alexander was worried about eating too much roast beef.

'Granny,' he said, 'if I eat all the meat, I won't have any room left for my chocolate pudding and I ordered one so I'll have to eat it.'

It didn't seem important that he had also ordered the main

course. Anyway, it all disappeared and the chocolate pudding went down in seconds!

I drive past Paible School most days. Soon they will start demolishing it. I can't understand why it's happening. A new building won't be the same good quality as the existing one. Maybe I'm just too thick to understand that it will save money in the long run. It would have been great to see the existing school being upgraded and the swimming pool in order once again. The gymnasium is beautiful and the classrooms are bright and roomy. They call it progress, the same as our Post Offices. It used to open five days a week and now it's two days, three hours a day.

October

Alexander didn't like Ellie going to university. He wanted her to be here with him. When she got a holiday job in Glasgow last year, he wasn't happy about that either. He adored Ellie just like she adored him.

Ellie got ill mid-December 2013. All the family were with her when she went through a major operation, and then received treatment. When the schools reopened, everyone came back home while Ellie and her mother returned to St Andrews so that Ellie could finish her final term, and receive treatment in Ninewells Hospital in Dundee. Since then, Alexander has had to be content with short visits which meant he spent very little time with Ellie and his mother. He was so brave and never complained.

The parents of his classmates were thoughtful and kind, taking him to their homes after school, and bringing him back after dinner. Then Claire came to look after him, and she was an absolute angel. She attended to mountains of paperwork too. With love and patience, Claire did so much towards filling the gaps in the young boy's life. With Alexander at the wheel, Claire would co-drive him down the beach or machair to pick up rubbish or take home driftwood. She played football with him, and he enjoyed helping her in the kitchen. Her 'job description' would be very hard to set out in detail! In the meantime, Sarah attended to her work with the horses in Wiltshire and, when possible, travelled up to visit Ellie and her mother.

I think the only time this year the whole family were together here was on Fraser's twenty-first birthday on 21st April. That was a very happy day and evening.

During all this time, Ellie studied diligently at St Andrews. Although she was receiving treatment, that did not deter her, and her loving mother was always at her side to make sure that she would eat and rest. Her father and Fraser would take short trips to visit her, and Ellie and her mother would come home for a few days. During this time, there was hope. The community here in North Uist and in many other places prayed for Ellie and the love and caring they showed this family was almost tangible.

Ellie sat her exams and gained a BSc Hons in Management in the Faculty of Science. Her graduation took place at St Andrews on 27th June. Once again, her family was together.`

Ellie was athletic, ran the hill race at the local sports in July 2013, was in the rowing club at St Andrews, and a few years ago with her student friends climbed Kilimanjaro.

However, in early August, her health deteriorated. During her second last trip home, her father took her and Alexander out in her kayak, and I'll never forget how happy she looked. The last month of her life was spent in Ninewells Hospital where again she had the constant love and care of her mother, by her side day and night. The rest of the family visited regularly, as did her student friends. I had precious hours with her myself, on the last Monday and Tuesday of her life when I travelled to Dundee. Alexander wanted to know what the last thing was that Ellie said to me. When I told him that we talked about him and the snowman, he was pleased because he had the happy memory of Ellie helping him to make it.

I prayed and pleaded with God to heal Ellie and not to take her away from her family, but that miracle did not happen. God gives, and God takes away, and I can't question his will. Her mother and father were by her bedside when she passed away peacefully at 5am on 18th September.

The funeral service was held very near her home in the Sollas Marquee. I don't know if you can call a funeral service beautiful, but many felt it was. It was a difficult task for Rev Iain MacAskill who conducted the service, and Rev Ewen Matheson who led in prayer. Both have been towers of strength in spiritual support for the family. Tributes were read and the singing was very moving. Well over five hundred people attended including a large number of local people, those who had travelled from other parts of the UK and others from the continent. Ellie's maternal grandparents came from France. Her heartbroken boyfriend, Robert, and his mother travelled from Poland.

After the internment, many people came back to the marquee where they could look at the photos displayed from the different times in Ellie's life. A wonderful group of local ladies had laid out a magnificent buffet, actually I should call it a banquet. Everything was donated, cooked and prepared by a loving community deep in sorrow. Donations towards Cancer Research amounted to over £3,500.

Ellie wanted people to wear bright colours, her own favourites being purple and lilac. Her blue kayak outside the marquee was full of purple heather and lavender which reminded us all of her happy times.

December

I heard about Sgoil Lionacleit[5] girls playing basketball recently and it made me think back to when I was attending Inverness Academy.

The only games played at Paible School were football and rounders. We didn't have P.E. in our curriculum and at playtime we organised our own games.

At Inverness we were taught how to play netball. I watched the goal shooter trying to get the ball in the net and they nearly always missed. The action was like throwing a sheaf of corn. I never liked using a hayfork to get the sheaves to the top of the corn stack and yet I could always throw the sheaf to the person on top of the corn stack. I asked the PE teacher if I could get a chance to be the shooter and she said yes. I never missed a goal.

'I thought you never played this game before,' she said. She couldn't understand how I could do this so well, and I was too shy to explain that I had plenty practice with corn sheaves. From that day on, I gained more confidence knowing I could do better than my classmates.

When I attend school functions at Paible, I'm always impressed with the children. They're so confident – my generation lacked that. I attended the Remembrance Service at Kilmuir Church and it was lovely to see the primary children taking part. It says a lot for the teachers and auxiliaries. It was well organised, and very moving. Paible, Lochmaddy and Carinish school children were all there and each child represented either a relative or someone born in their village. I felt proud and emotional when the children paraded, naturally more so when Alexander carried his wooden cross with the poppy on it and respectfully spoke in Gaidhlig, 'I am Donald MacDonald, Kyles.'

That Donald was a second cousin of Archie MacDonald, my father, and Alexander's great grandfather. Donald and his brothers, John and Alan, all fought in the war. Sadly, Donald and John were killed but Alan returned. Alan's grandchildren live in Kyles and in many other places.

I wonder if you saw the TV programme by David Attenborough about the barnacle geese nesting on rocky cliffs 400 ft high? The day after hatching they actually jump down to ground level to reach the grass. What a feat! It was amazing to watch. I think three out of four goslings survived the jump. I was nearly in tears watching as the goose and gander walked along proudly with their young before a fox came along and devoured them. I guess it is survival of the fittest!

2015

February

Thursday 8th January was a night we won't forget in a hurry. I believe the wind strength was about 103 mph, particularly frightening so close to the anniversary of 11th January 2005 when gusts of 145 mph were recorded in Benbecula. It was a terrifying night with lightning and thunder. For many of us, the power went off at 2 am. I didn't dare go to my bed. Thankfully, no lives were lost, and the considerable damage done to houses and sheds can be repaired. But my dogs were so scared, panting and shaking, and wouldn't move away from me. There was bright moonlight so I could see the hen and chicken house. Surprisingly, none of them blew away or were damaged.

Friday was better but many homes were still without power. It wasn't a big problem for those of us with solid-fuel stoves. I let the dogs out about 8.30 pm, but Cola did not return. I assumed she had gone next door so I locked up and went to bed. At 3 am, I was woken by a cold nose on my cheek. It was Rocky who had never been in my bedroom before.

'Rocky, get away,' I said.

But he whimpered and clearly wasn't going away until I got up. He took me to the door, stood looking expectantly and when I opened the door, there was Cola, not even looking guilty for waking me. She went straight to her bed and Rocky went back to his, looking quite proud of himself.

The wind had dropped from the previous night and in bright moonlight, there was the Hydro team sorting the problem in the middle of the night. Their work is dangerous and we should never take them for granted. I watched until 3.30 am when the big light went off and they drove away in their vehicles. Shortly after that, the lights came on.

March

I watched a TV programme called Animals in Love that interested me. Working with sheep and cattle most of my life, I believe that mating has nothing to do with love, but when it comes to lambing and calving, there are powerful bonds. Cows will do anything possible to protect their young. I would definitely call family, fold and flock bonds of love. When calves are reared in batches – if they are allowed to – they will form lifetime bonds and become quite miserable if separated. I have watched flocks of sheep being shepherded to fresh grazing and mixed up with other flocks, but then splitting up and staying in their own flock. It's really amazing how they know where they belong.

In 1985, we bought a Highland bull and a heifer in Oban and they both travelled to Lochboisdale haltered in a pen on the deck. We had that bull for three years and during his stay here, despite serving the other cows, he always came back and stayed beside his travelling companion.

I remember back in the mid 50s when we had horses, and we had a mother and daughter. In those days, the only feeding was sheaves of corn. The horses were side by side in a stable and when I put a couple of corn sheaves in the manger, the mare would bury her head in the corn and throw all she could towards the filly.

A cow isn't that kind hearted. Put some feed like barley or nuts in a trough for a cow and calf and the cow will eat as much as she can, even if the calf gets nothing. She doesn't drive it away, naturally she loves to let it suckle her, but when it comes to tasty bites, she is very selfish.

Now, poultry has strange ways. When you put fresh food on the ground, the cockerel will get there first, but will hardly eat anything until he gathers the hens by calling them over. The broody hens will make sure that their chickens get fed first. Similar to mammals, hens have their pals and at night you find the same hens perched side by side.

In 2013, Muran, a yellow heifer, should have had her first calf but she was empty. However, in 2014, she calved and produced twins. One was a red male but unfortunately the other was a black female, and it's a fact that a female whose twin is a male is very unlikely to breed. Such a beast is called a 'freemartin'.

Last Wednesday, the weather was foul with wind and heavy

143

rain, and all remaining calves were to be taken off Vallay. The plan was to house the females and leave the males outside. I kept wondering about the twins and that it would be a shame to separate them. So, should the female go with the males or the male come inside with the females? However, nature made the decision. The male was cold and didn't look too good, so a small pen was made for him beside the females. The females could go in and out of their large pen. They didn't cry too much for their mothers and seemed happy having a warm shelter inside. I watched the black female twin who stayed in and ate the hay that was right beside her male twin. She wasn't going to leave her brother.

I was listening to Coinneach MacIomhair's chat show on Radio nan Gaidheal. Two young crofters from Uist, Angus Ferguson and Iain Stephen Morrison, and one from Lewis, talked about how difficult it was for young crofters to get started. There are numerous problems and the government is not doing enough to support them. One long-standing problem is non-active crofters living on crofts but not allowing anyone else to use them. I'm sure those young men could have talked all day and I was delighted that there are young crofters who will raise their voices. Well done, lads!

May

At the end of March I travelled to Assynt to attend a young crofters gathering at Glen Canisp Lodge organised by the Scottish Crofting Federation. I was privileged to travel with Sue Dancey and her Crofting Year class from Sgoil Lionacleit. With other 'old' crofters, I was invited to talk to the young ones and to answer their many questions. It was a treat to listen to ninety, enthusiastic young people. The children from Sgoil Lionacleit mixed very well and I was proud of them.

The young crofters decided to set up their own group to lobby the government and give crofting a stronger voice. It is refreshing that a new generation will represent crofting, and is willing to fight for changes in policy.

We enjoyed a lively ceilidh on the Friday night with Margaret Bennett as MC who sings beautifully, and is a mine of knowledge about the past. The young ones also sang, played the accordion, recited poetry and told stories, all very traditional and most enjoyable. Dinner was a whole pig, roasted on a spit. What a treat!

Last year we were still seeing barnacle geese as late as mid May. Of course the greylag geese never migrate now. Yesterday in Vallay, a goose flew up out of the bent grass, and I found five eggs in a nest. I left one and took the rest home to make a lovely omelette.

Work has started on the new school and it makes me sad. If buildings could talk, I imagined what the old and new schools would say to each other.

Paible Junior Secondary[1]:
'I was built in 1904. There was no electricity then. The children carried peat to school to burn on the open fires and would not dare misbehave. They qualified as doctors, nurses, professors, teachers, solicitors and a Squadron Leader. Look at you across the bay, built in 1963 and about to end up as rubble. You had everything – a beautiful gymnasium, swimming pool and central heating, and now you are not good enough to accommodate the few primary children. When they were building you, a storm blew in May and you collapsed. At my ripe old age, I even withstood the 145 mph hurricane of 2005. There is money around now, and it is normal and fashionable to waste it.'

The 1963 school would likely listen and say very little:

I don't want to disappear. I could have been around for as long as you but I have to give up because I'm 'not fit for purpose'. That is a phrase that is used a lot these days. I am sad that the majority of parents and teachers want to see the end of me. I hope my replacement will be as robust as I was, and maybe in the next decade there will be a huge increase in population and then the policy makers will wish they hadn't got rid of me'.

Back in the late 40s, we took a packed lunch to school of homemade scones and oatcakes, and maybe rock buns, a dumpling and a bottle of milk. One of my classmates always had plain loaf sandwiches with the crust cut off. It was a novelty for me to eat loaf. My mother gave us scone sandwiches with homemade butter, sometimes with crowdie, cheese, hardboiled eggs chopped and mixed with fresh cream, or poached sea trout. What luxuries we had! Yet I loved to exchange my scones with the shop bread.

On a dreadful day in March, a young cow, Fraoch Odhar, gave birth to a premature black female calf. We found it wandering about in Vallay sucking from a cow that hadn't calved but of course, she didn't have milk. Then young Fraoch came along but she didn't have a lot of milk either. She took the calf away to shelter, but she wasn't as protective as you would expect. I mixed a feed of artificial colostrum and warm water, and the young calf willingly sucked it. The cow only allows the calf to suck her from behind, and never licks it. Ruth and I bring her nuts in an old brown pillowcase so that the other cows don't hear a noise and come running. When we arrive, Fraoch recognises the jeep and calls the calf. She knows she won't get the goodies unless the calf is there. Both follow us into a secluded hollow, and as soon as we put the nuts on the ground, the calf starts sucking. I give her a bottle of milk too as a supplement. Fraoch's milk supply is now increasing, and the calf is thriving, although it is small and we wish the mother would give it some TLC.

June

I was listening to a radio programme about the sea eagle. If I was a young sheep farmer, I would be gathering a crowd and marching to Holyrood. That word 'managing' is great. Scottish Natural Heritage (SNH) and the Royal Society for Protection for Birds (RSPB) must have magical powers if they can 'manage' the most powerful bird in the sky. British standards for animal welfare are supposed to be the best and yet tourists who feast their eyes upon the eagles while knowing how vicious they are, would be the first to report a crofter or farmer for supposedly neglecting a lamb. Seemingly, sea eagles have left the island of Rum but as far as I know, there are no sheep. The conservationists who are 'managing' are smiling all the way to the bank with full-time jobs for many years but the crofter or farmer gets no compensation for losses. If a dog is caught attacking sheep, he is shot, but if an eagle is seen attacking a lamb or calf, you have to let it get on with it.

The most beautiful weather this year was on the day of the North Uist Ploughing Match. There were plenty of competitors and a huge crowd gathered to watch. The usual well-organised Sollas community spirit shone through, and plenty of homemade food was available. It was great to see people young and old taking part with tractors and working together. When I was a child, ploughing was done by horse-drawn ploughs and the last ploughing match was held in 1939. Mary Ann MacDonald attended as a young child, and was delighted to be there in 2015. The judge had a difficult job but I think his decision was unanimous. His wife had the hardest task, judging the best looking ploughman! I hope this enjoyable event will continue every spring for many years to come.

When I was at school, we used to run to Horisary strand, near Bayhead, to collect cockles during our lunch break. Today, school children don't need to sacrifice their lunch break because they go on walks when the weather is warm. It would be nice for them to throw off their shoes and bring a kitchen fork and plastic jug, and be shown how to find cockles.

July

Becky's Dream Cow

The Robson family lived for five years in Kyles, just next door to us. Becky and her sisters, Geraldine and Michelle, had a happy childhood and loved to come with me to help look after the animals. They had their own dogs, cats and pet lambs. I still remember the ewe lamb, Milly Molly Mandy, and the wedder[2], Rupert, who lived to be a teenager, recognised his name and didn't land on anyone's plate!

Geraldine loved feeding my pet ram lamb, Maradona. She took him to the North Uist Show in a dog-collar and lead. Kenny MacKenzie was the judge and when he asked her the name of her pet, she proudly called out Maradona! Everyone laughed, and Kenny decided that he deserved first prize, and remarked, 'I don't think anything is getting past Maradona this year!' It was World Cup time.

After the family left North Uist in 1990, we kept in touch. They got a croft in Holm, Stornoway, and the children were delighted. Next to them lived George Morrison, a butcher, who bought fatstock[3] in Dingwall to slaughter in Stornoway.

One evening 14-year-old Becky phoned me. 'Ena, when are you coming up?'

There was something excited and urgent in Becky's voice so I listened carefully.

'Ena, I was looking at George Morrison's cattle, and there is a Highlander among them. Maybe I could buy her and have my own dream cow! Oh, Ena, she is lovely and I don't want him to kill her. Will you come and talk to him?'

The next day I went to Stornoway and Becky showed me the heifer. She was lovely, and though dehorned, had a lot of character. The more I looked at her, the more I thought she resembled some of our own Highland cattle. Cattle, like humans, look like their parents. George Morrison told me he bought her in Dingwall from Bob Shaw, who was a butcher and cattle dealer. Bob Shaw was a regular at the Uist sales and had bought this heifer in Lochmaddy. I asked George for her tag number, and when she was going to be slaughtered. He kindly gave me the details and promised he would not do anything until he heard from me again.

Becky was getting more excited by the minute. I can't remember if she had even told her parents what was going on. We went back to her parents' house. Knowing her tag number and that she came from North Uist, I had to do some research. I phoned Benbecula Agricultural Office and they told me the tag number belonged to the brothers Ruairidh Archie and Iain Hector from Knockquien, Carinish.

I had to know about her background. If she was a twin to a male, only five per cent of them can breed. Or there could be a health problem. However, a phone call to the MacDonald brothers solved everything. They were amazed about the coincidences, and so was I. The brothers had bought a Highland bull called Sradag from us, an old-fashioned brindle bull. His mother was Lasair Chlach Chanaidh, Flint Stone of Canna, so I called the bull Sradag, Gàidhlig for the spark that comes from a flint stone. I had bought her in Oban in 1979, as a three-year-old from Margaret Fay Shaw, Canna. She was the only black Highlander at the sale because black had gone out of fashion! And, indeed, Sradag turned out to be the father of this heifer. No wonder I was seeing a resemblance.

I was just as excited as Becky to discover her background. Then it was down to the nitty-gritty. Mum and Dad would have to share her dream with their permission to buy the heifer, as well as some funding. I had to phone George Morrison and ask for a price. He was a real gentleman and I think he asked the same price as he had paid for her. The whole experience gave him a lot of pleasure.

Becky went into action as soon as she had her parents' permission. She is independent and wanted to use her own savings. It was so comic. She came downstairs with her piggy banks and her other savings, money she had received at birthdays and Christmas. All of it was poured on the table, all to save the heifer's life. Then she had to decide on a name. It didn't take her long. Aisling which is Gàidhlig for Dream.

Aisling spent her first year with an Icelandic pony and Milly Molly Mandy and Rupert for company. What a happy bunch. Aisling had many calves, there's a small fold of Highlanders descended from her. She lived a very long and happy life, and I am sure she was the most treasured cow in the Western Isles.

With all her care and commitment to so many animals over her young years, I always thought Becky might become a vet. But, as it turned out, she trained as a pharmacist, and has subsequently taken further training to be a doctor.

August

North Uist Highland Games or Sports as we used to call that special day, was a real treat on a beautiful sunny day. It was good to meet many old friends visiting on their annual holidays. As I walked, or should I say talked, my way around the sports ring, I was thinking back.

As children, we walked the five miles from Kyles to Hosta. Everyone did. There were very few cars in North Uist then, although I remember once getting a lift in the back of Angus MacLellan's lorry. I can't remember if there was any food but we were fortunate to have Auntie Annie across the road in Hosta. Talk of an open house! Auntie Annie and Gilleasbuig lived in a thatched house, beautifully kept, with a vegetable garden in front surrounded by garden flowers. More than half the people at the sports were invited to the house, not just for a cup of tea, but a complete dinner, all home made and produced. There wasn't any electricity and the only cooking facility was a Victoress Peat Stove. She would have delicious mutton soup, roast mutton, carrots, cabbage and other vegetables from their garden, new machair potatoes, and of course gravy that nowadays chefs don't seem able to make. Then *Carrageen*, seaweed pudding, or semolina pudding, both topped with stewed rhubarb and smothered with real cream. How did everyone fit into that house?

Back then, there were very few Highland dancers because there were no dancing teachers. But there were two girls who were dancers, Chrissie Dick and Rena Harvey, who lived on the mainland. How I envied them in their beautiful kilt outfits. When I went home, I used to stand in front of the wardrobe mirror and pretend to dance the Highland Fling!

In 2013, in Loch Carnan, South Uist, Peter Beaton had a lovely Highland cow due to calf. I believe she calved but then had a prolapse. After the vet treated her, she could not get up. Peter loves his animals and although there was not much hope for the cow, he nursed her for more than three weeks, lifting her and giving her every possible care. Peter's patience must be admired for the cow eventually got up and started grazing normally. Earlier this year, the cow and two bullocks arrived at Ardbhan, destined for slaughter. The cow was empty because another calf could have been too much for her. I saw her often, usually a distance away from the rest. She had a lovely head with wide horns, very appealing. She was not old, and as I got fonder of her, I felt sad that she would have to be slaughtered. She was heavy for a non-pregnant cow but her good condition was because she hadn't reared a calf for two years. When we were busy attending to the other cows, we didn't take much heed of her. We would see her with the binoculars, grazing in the distance. Then, one day, I was standing beside her and looked on in disbelief – the cow was starting to make an udder and was definitely in calf. She was taken home to Kyles where we could keep an eye on her. I spoke to Geert, the vet, to check if she was likely to have another prolapse and he assured me that there wasn't a huge risk, but to keep an eye on her. Despite doing just that, she calved when we were all in bed! In the morning we found a lovely red female Highlander sucking a proud mother, its tail covered with 'yellow toffee' confirmed everything was in good working order. Peter was thrilled with the good news. It was an exciting royal birth!

September

Since October last year, we have had dreadful weather – the worst I have experienced, but I do remember other difficult times. In the 1940s, during storms, corn stacks blew away into the sea because they were only secured with corn ropes. Later on, discarded fishing nets would land on the beach that proved excellent for securing them. In 1964, we were still trying to make corn ricks, (*toitean*) in November, after we had to move the stooks (*suidheachain*) out of the water many times. Later on, I think it was in 1968, we planted the potatoes in April and then, when the shaws were about six inches high, we got a hard frost that killed them off. In 1985 we experienced a very wet summer and struggled to get any dry hay. Sometime in the 1970s, we had an exceptionally dry summer and most of the machair corn was so short it could not be cut. I can think of many hard times and crofters back then didn't have money to buy feeding from the mainland. Nowadays, crofters have other jobs and they are usually not as hard up as before.

There is talk of rewilding parts of the country by reintroducing wolves and lynx. It is bad enough having eagles, ravens and geese! I don't like to hear all these rumblings because where there's smoke, there's fire and once it happens, there's no turning back. However, it benefits those project workers who have a lifetime job monitoring progress and making sure that livestock deaths had nothing to do with their pets. Who would imagine fifty years ago that mink, ferrets, hedgehogs and eagles would be enjoying living in North Uist?

There are a lot of tourists and relatives from the mainland here this summer. Certainly, the roads have been very busy. It's good to see young families coming to the islands. Two years ago, a young family stayed in Kyles and Rocky had a great time with the children. They came back this year and borrowed Rocky. He was in and out of the sea, and ran up and down the beach. The family called before they left and brought me a beautiful lemon cake as a thank you. No wonder Rocky's been losing weight

152

with all the running around. Maybe I could start a summer business in my old age: Rent-a-dog!

Fraoch, the premature Highland black calf is thriving well. She is very small for her age, but at last is in normal condition. I had to take her away from her mother – they had a strange relationship. I could see that the mother didn't really care for the calf, and her milk never increased. I took Fraoch home for two days to see what was happening. When I took the calf back, its mother never even welcomed it. I tried her udder and it was almost empty. She allowed the calf to suck but she didn't lick it, not even once, and as soon as she was finished eating she walked away, as though she couldn't care less. I felt sorry for the wee pet as she stood there watching her mother walk away. I almost cried. But as she was used to a bottle, she came running over to me. So we put her in the back of the jeep and now she gets all the attention she wants. I know she's happy and she has a pal, an orphaned female.

October

I'm sure many people feel as I do – when we enjoy eating something delicious we wish we could share it with those who are hungry.

In the 40s and 50s, we ate so many rabbits; most people did. Mammy, as we called her, was a good cook. And it wasn't just the cooking she had to do, it was also the gutting and skinning, including the head which was boiled for stock, and the rest prepared in various ways. If the rabbit was old, it was difficult to skin and it was boiled and served with white parsley sauce. Or she would have boiled chunks of carrot, turnip, and whole onions, along with the rabbit so all the vegetable juices were in the stock. Sometimes she fried the boiled rabbit and potatoes together. A younger rabbit would be fat and browned in an iron pot on top of the Modern Mistress stove. It would then be stewed slowly along with plenty of our vegetables. When very young it was pot-roasted with butter and that was tastier than a young chicken.

Some rabbits could not be cooked – those in milk or pregnant and, thinking back, I am sure there were times when my mother was glad to give a rabbit to the dogs or cats. She must have been very tired of all the preparation, and having to cater for school lodgers as well as the five of us. I was scunnered[4] with rabbit when myxomatosis came along because it was an ugly disease. It has now disappeared. I was given a rabbit, a young fully grown one, last week by someone special. I cooked it as I was taught by my mother, adding the vegetables that are growing in my garden, and managed to get three dinners out of it. Of course I had potatoes too. My generation does not enjoy dinner without potatoes.

There was an interesting article in the Scottish Farmer. A group of farmers from Norway came to visit farmers in this country and warned of the consequences of so-called re-wilding with lynx and wolves, as well as the sea eagles we already have. Last year in Norway, these predators killed 3,895 ewes and 19,671 lambs. It is time that shortsighted and possibly criminal people are confronted with reality and put in their place, preferably on a desert island with those predators for company.

I look after four pet calves, one of which is Fraoch, the premature calf. The other Highland is the only male, the two other females being Highland X Shorthorn. I wonder if the Highland breed is more intelligent than others? I have been penning them in small paddocks by the house. One day I put the four of them in a two-acre field which had been cut early in August for silage. I put their trough near the gate, and when I went to feed them, they were at the far end. Rocky is always with me, and when I called the calves, wee Fraoch and the male Highlander came running, but the other two stood there looking confused. So I told Rocky to go and get Torloiska and he duly brought them to their feed. He knows their names. Next day they all came together. We don't take Rocky among the cows and calves. It would be dangerous among protective mothers.

Some of you may remember me writing about Serge, the agricultural student who came to us seven years ago for practical experience. This year he phoned in the spring, wondering whether he could come and help for two weeks in August. It could not have been better timing. He is so capable. He varnished my wooden doors, cut firewood, and helped weed the garden. It is amazing that a 25 year-old would come voluntarily from Luxembourg to do all that.

November

I have just returned from an enjoyable twelve-day break to the mainland, visiting family and friends between Glasgow and Ayr. Throughout the trip, the weather was good with lots of people walking around in short sleeves as if it was July.

September and the first part of October were great, so now the wind and rain is particularly unwelcome. Maybe the bad weather took a break like I did, but then decided it felt more at home in the north west of Scotland. It was good to forget about wellingtons and having to feed cats, dogs, hens and pet calves … this year at least there are no surprise chickens.

Those small animals were well catered for during my absence. My young nephew's family enjoyed helping their dad, Colin, feed the hens and dogs. However, the youngest one, Alastair, found Rocky a bit boisterous, particularly because when Rocky is standing on his hind legs, he and Alastair are the same height!

While I was away, I attended the Highland Cattle sale in Oban. Although the quality of the cattle was good, I am afraid that prices were on the whole disappointing. This year's weaned heifer calves did command good prices, but the three-year-old in-calf heifers did not.

I was glad to hear that the cattle sale at Lochmaddy on 23rd October was good, with prices comparing favourably with the mainland. Because the spring and summer grazing was so poor, cattle were lighter this season, but nevertheless the trade was better than last year. I was also pleased to hear that John Allan MacLellan's calf won the show before the sale.

At least the calf premium is continuing. It was the Scottish Crofting Federation (SCF) that fought for that while the National Farmers Union (NFU) was against the payment. When there was talk of abolishing the premium, the NFU was up in arms, as though they had introduced it in the first place. The SCF seriously needs strong, young voices to lobby and represent the views of crofters. The way I see it, CAP reform, and other schemes, are becoming more complicated and difficult to access.

I am aware that a goose count took place recently, but I am too angry to write about it at the moment. From the numbers quoted after the count, you would think they were counting eagles, not geese. Many of us regularly see flocks of a thousand or more and they are most certainly not fewer in number than previous years. I am now very much aware that there are heavy losses of lambs occurring, either from the sea eagle or golden eagle or both.

For all the problems and difficulties we have in these islands, we are still so much better off than the city dwellers. Not only do we have the natural world at our doorstep, we also have a community where people care for one another, sharing joys and sadness alike. Much as I enjoy my visits to the mainland, I am always so glad to come home.

December
Global warming! If we didn't have access to news coverage, an Outer Hebrides climate change summit would be discussing where our summers have gone! Mind you, I wouldn't like it as hot and dry as many years ago with parched and cracked soil. Before

there was running water, some of the wells would dry up and some of us had to walk a long way for a bucket of water. I can remember my mother trying to make butter with the wooden upright churn standing in a large basin full of cold water to keep it cool. Cream for butter making cannot be too cold or too warm. The flies were also a menace, and we had to avoid getting sunburnt.

Something quite interesting on the news recently was a piece about farming insects and feeding maggots to pigs and poultry to provide them with protein! I have very little knowledge of pigs, but I know they are not very fussy about their diet. Poultry will also eat almost anything. There is nothing new in seeing hens or other birds enjoying a feed of insects. Turn over a stone in heavy soil and underneath you will find earwigs or earthworms and just watch the hens' excitement – not many of the insects will escape! I just hope they don't start breeding maggots here and then abandoning them. Next, we will have a plague of blue bottle flies attacking our sheep.

We all have our favourite jumpers and jackets, and many of my age group grew up having to look after our smart clothes and did that so carefully that they lasted, might I say, way past their wear by date. I have been as guilty as anyone at holding on to clothes. I go through them and decide they are too old fashioned for the charity shop, but too good for recycling. Or, I remember that an old aunt or big sister knitted that cardigan. Or that skirt is a bit tight, but maybe if I lost an inch around the waist, it could be worn again. So back it all goes into the wardrobe. Then, after all the attempts at a clearance, there is nothing left to take to the charity shop.

Anyway, recently, there was a notice in the shop that read 'warm clothes wanted for Syrian refugees in Serbia'. Well this was comparable to finding a good home for kittens! No decisions had to be made – at last someone was going to be very happy to get warm clothes. If I had to live in a tent in atrocious weather, the last thing I would look at would be a designer label. I think I would be grabbing the warmest coat whether it was two or twenty years old. I'm now so thankful that I didn't part with those old clothes. I spent an enjoyable couple of hours neatly folding the clothes into bin bags and took them to Taigh Chearsabhagh in Lochmaddy from where they will be delivered to Serbia via Glasgow.

During November, in Paible School, the North Uist branch of the Scottish Crofting Federation held its AGM. We had a great meeting and it was a tonic to see so many young crofters attend. We have formed a new committee of young, active, intelligent crofters who will protect the future of crofting. They have a lot of work to do and will need all the support of the older ones.

My pet calves are thriving. Fraoch – the premature calf – is now almost as big as the rest. It's hard to believe that I could lift her into the back of my jeep as though she was a heavy shopping bag! She is in great condition and very cute. We had lots of problems this summer with the cows and the bull paddling across the lochs to eat the older thick grass around Ard Heisgeir, near the main road. We never used to have those problems. Maybe it was the goose pollution that drove them out. Geese don't like long grass, so the roadside grass would have been tastier.

2016

February

I wonder why storms have names now. I thought I would remember them, but I don't, maybe because I want the storms to move on quickly. A friend had to remind me: Abigail, Barney, Clodagh and Desmond. If the Western Isles Met Office named the storms, we would be three quarters down the alphabet and at Tornado by now! When does a severe gale become a storm? Most of the time the forecasters get it right, but waiting on the storm can be as stressful as experiencing it. And the ferries have been much disrupted. We hear about rainfall records but not so much about wind speed. Does anybody in North Uist keep a wind speed record? It has been a frightening winter.

In June 1987, I took a black Highland bull, Pibroch Dubh, to the Royal Highland Show. He travelled in a pen on the open deck of MV Lord of the Isles from Lochboisdale to Oban where I was met by the late Mary MacLean from Benderloch. I stayed the night on her farm, and the next day she and her brothers and I travelled in Archie MacPherson's big cattle float to Ingliston. Pibroch travelled in style with Mary's beautiful blonde females that won many prizes, and the MacPherson cross-bred bullocks that were champions.

Even if I was as strong and able as I was back then, can those of you who want to see the end of the route Lochboisdale to Oban tell me how I would get to Oban? I should sail to Mallaig, and the animal would be in a float, but remember, I don't want to drive on the mainland so someone would have to drive from Oban to Mallaig to meet me. Surely that is very inconvenient.

Now, imagine a crofter has to get more than 30 head of cattle to Oban. The crofter books a local artic[1] lorry but it's too high to get under the bridges on the road to Mallaig. And he has to get to Barra to catch the ferry to Oban. Anyway, bookings are made, and the lorry is driven fifty miles to Eriskay while the weather deteriorates, so the Eriskay to Barra ferry is cancelled.

United Auctions has a good lairage[2] facility in Oban. Crofters in Uist find it convenient when they are buying in orphaned

alves, sheep and bulls. If the ferry is late, you know the stock
s well looked after and United Auctions will see that they get
o Lochboisdale, when the time is right. What will happen to
his important service?

March
I much prefer radio to television and I listen to the early Saturday
show Out of Doors with Mark Steven and Ewen McIllwraith.
It was an extra special show on Saturday 13th February when
they broadcast from Canna and spoke to a friend of mine,
Geraldine MacKinnon, the farm manager. She appeared natural
and relaxed. I have been to Canna twice, so I could picture what
she talked about. Canna belongs to the National Trust for
Scotland (NTS) and they have had thousands of rabbits munching
away all the grass. NTS paid trappers to eradicate most of them
and then left one trapper there all year round to control what
remained of the population. That's what we need for goose
control here.

Geraldine had similar views to myself about grazing restrictions.
Canna has many corncrakes, but Geraldine disagrees with RSPB
about leaving grazing undisturbed until August or September.
The grass grows too tall and thick, and the birds can't walk
easily amongst it. I wish the policy makers would listen to us
and realise that livestock are at a loss.

Modern technology is sheer magic to someone like me. The
young, and not so young, tell me what they can do, and it's like
a fairytale.

At the end of January, when they were feeding silage on Vallay,
the 2014 heifers were running fast behind the tractor with Seonaid
Casfhionn in the lead. Casfhionn is Gaidhlig for White Feet. She
is a red pedigree Highlander with shaggy white feet and a white
patch on her forehead whom Ruth and I nicknamed The
Clydesdale. Fraser took a photo of the heifers running and put
it on Facebook. That evening he got an email from Volker in

Germany asking if he could buy Seonaid. He liked the white socks. And so Seonaid was taken from Vallay, but she couldn't join the other five Oban-bound heifers until she was tested. The only empty shed available was Sarah's stable.

Geert, the vet, came the next day and after eight days the negative results arrived. Soon after, Seonaid was off to Oban and now she is in Dumfriesshire at a quarantine farm for thirty days when she will be retested, and then transported to Germany. The five calves that were auctioned in Oban are all going abroad as well. Volker and another German bought duns so those three will travel together. Two are going to Italy, and my own black favourite – whose grandmother came from Canna – will travel to Denmark. A red two-and-a-half year old bull was sold to a couple in Devon.

Volker, the German buyer, came over to Kyles for a couple of days after the sale. He was very interested in *A Keen Eye* by Una Cochrane. Una has researched Highland Cattle and explains, with photos, that the casfhionn[3] is not a fault. Una gifted me this book and the second edition is now worth £1000.

I have three cats and I call them all Cissy. I called them to feed and only the young ones appeared. Then the mother cat came running out of the shed with a fat young rat in her mouth. She was very hesitant, probably trying to decide if she was going to eat the rat or join the family dinner. She dropped the rat, quickly ate the food, determined to get her share of the dinner, and then ran back to the dead rat, picked it up and disappeared into the shed.

May

At the end of March, my sister Agnes passed away. I have many happy memories of her and I feel prompted to share them with my regular readers.

Agnes loved reading my articles in Am Paipear, so for years I had the paper delivered to her monthly. I don't think any of my readers looked forward to it more than Agnes. The following Sunday, she would take it to the Free Church in Ayr where it was passed round before being returned. She would then cut out my column and stick it in her scrapbook. When I was tired, busy, or struggling to find something to write about, I forced myself to write something interesting so as not to disappoint Agnes. Her favourite story was the one about Alexander and the snowman. I think that is my favourite as well.

Agnes loved cooking and baking and sharing what she made with old neighbours and those in need. Agnes wanted to make some white puddings, but while she could get suet from the local butcher, skins were never available and she didn't want to use a plastic skin. When we were discussing this on the phone, I remembered that I had some skins washed and ready in my freezer. I had been making some *maragan*[4] in November and had put some aside for future use. Donna Nicholson from Ayr was home at New Year and she kindly delivered the skins to Agnes. The whole family loved them and knowing Agnes, I don't suppose she saw much of them herself!

About six years ago, Agnes's daughter, Kirsty, Ross, and young John came in their camper van. On the day they were leaving, I went out to the freezer in the shed to get them some mince to take home. When I opened the freezer, I couldn't believe what I saw! Someone had used the socket to plug in an implement and forgotten to plug the freezer back in. The whole top layer was defrosted! It was more than I could cope with, so Kirsty took the rest in a large cool bag and Agnes received all the mince later that evening. She made large and small mince pies and shepherd's pies which were all frozen, but only after everyone on Teviot Street got a mince treat that day! That was the way she was, she liked to give. She was always making crochet blankets which were beautiful, especially those for new babies. I don't know how she found the time because she was always happily feeding her family, her grandchildren, and latterly her great-grandchildren.

Agnes was three-and-a-half years older than me, and often

reminded me that when I started school, we were in the same classroom. However, when Agnes was in P3, she started learning sewing and knitting in a different room. As soon as she left the room, I started screaming and howling! In the end, the teacher relented and let her take me with her. I'm sure I was an embarrassment to her. When she reminded me of the incident, she admitted saying, 'That rotten girl. She's totally spoilt!'

My two sisters and I will always miss her dearly, but the break is harder for her widowed husband and family.

Our calving is half way with lots of lambs around. It's very cold for them but better than constant rain, I suppose.

On Saturday we came across a newly born calf, a large strong female which is proving rare this season. There were five cows near the calf, none the mother. We searched the area but didn't find a newly calved cow. We gave the calf a warm drink and left it with the cows. At low tide on Sunday, we couldn't find the calf, but Fraser came across an obviously newly calved cow so now we had the mother but no calf. On Monday, John MacBain and I went over and again saw the cow but not the calf. We went back again on Tuesday. On Monday, we'd fed silage so the cows were all in the same area and we spotted the calf wandering among them, looking for a foster mother who was happy to let it suck. Then we spotted the cow and took the calf in the jeep to her. The cow looked but didn't respond. Other cows licked the calf and made it welcome, but Seonaid was not interested. We took the calf home, I gave it milk and put it in a warm stable. On Wednesday, the cow was brought home. Norman made a pallet pen beside a large pen. The cow was brought into the big pen where she could touch and smell the calf through the bars. On Thursday morning, Fraser put the calf in with the cow and returned the calf to the pallet pen once it had sucked enough. On Friday, the mother made a great fuss of her calf, even protecting it, which was a natural thing to do. So, with these good results, she was driven back to Vallay and we hope all will be well. I wonder why the calf left in the first place. Did she have a difficult birth? It's rare in Highlanders, mind you. We will never know.

June

Many of you would have seen some of the Queen's 90th birthday celebrations on TV. It was spectacular with some of the most beautiful horses in the world being ridden by first class riders in Windsor Park. It must take years of skill and patience to train man and horse to that gold star standard. The highlight of the show was seeing the Australian horse whisperer working with ten horses. They weren't being ridden and they didn't even have bridles, yet he got them to parade in formation. Horses have brilliant brains. I wish cattle could be led like that!

Recently we had a family celebration with a romantic and royal flavour. The Queen's birthday is on 21st April and my grandson, Fraser, and his girlfriend, Carianne, both share the same birthday. Fraser was 23 and Carianne was 22. As if it wasn't enough to celebrate right in the middle of lambing and calving, a week before the double birthday, the romance began. A small package was hidden in my bedroom, a tight secret between Fraser and Granny. On 21st April both himself and Carianne attended to the lambing and checked the cows and calves as usual. In the evening they had to go to Griminish to shut a gate that some thoughtless person had left open. From that spot, on a clear calm evening, there is a panoramic view with St Kilda impressive on the horizon. It's one of Carianne's favourite views and I believe Fraser decided this was the right time and place. A little bird told me that Fraser made Carianne feel like his queen by proposing on his knees on the heather moor and giving her a beautiful engagement ring! Who says crofters are not romantic? I wish them every blessing and happiness.

The new Paible School is progressing right on time which means the old one will be demolished soon. I still think it was wrong to build a new school because it is out of place in Paible, has no character and looks like something you would buy from IKEA. I hear that the number of children is increasing so maybe in five years time they will need to build extensions!

When I was in my final year at Paible School, Class 3, S3 nowadays, we had to sit two exams, one called Control with the result determining whether you attended an academy or a technical school in Class 4. Our other exam was organised by the Church of Scotland and was a biblical test. We were told which books we had to study. That year it was 2nd Samuel and the prize was £20 which was a lot of money.

We did a lot of Bible study in school and memorised many passages. I remember walking to school barefoot with buttercups stuck between our toes. One of us would carry the Fible and we revised the verses which we would have to recite in school. We felt it was a dull subject but I think we were privileged to have learned so much. Those early teachings helped to guide us through life.

All of us in the Free Church are looking forward to having a resident minister soon. Rev Calum Murdo Smith will be inducted on Saturday 18th June. He has a family of three so that will help the school roll and hopefully encourage young people to attend church.

The EU membership referendum is fast approaching. I think we should stay in the EU. I remember in the early 1980s how we waited to hear if Objective One money[4] was coming to the Western Isles. Through powerful lobbying, the late Winnie Ewing achieved her goal and we received £20 million. That was the Integrated Development Programme which we called the IDP and nicknamed 'I Don't Pay'. Another Scottish referendum is the last thing I want.

August

Alexander is eleven and enjoying the holidays with other young boys his age. Earlier in the summer he attended *Feis Tir an Eorna*[5] and received tuition in the *chanter*[6] and shinty[7]. I am very happy he has a great interest in both activities. He won a beautiful silver-mounted chanter in a raffle, I think, and that in itself is encouraging him.

Alexander and his dad caught four rabbits a few days ago. He came to my door at 10 pm with orders to stew them! He sounded like my father. He would bring plenty food home and poor mother had to do the skinning, preparation and cooking. I cooked the rabbits in butter and they were a treat along with our own new potatoes. We are very well off on this island, with an abundance of shellfish, fish, meat and healthy vegetables.

North Uist Agricultural Show was a great success with very good weather and a record number of people in attendance. The judge, John Scott, was very impressed with the livestock. It is good to see so many young people on the committee and indeed they must be congratulated for all their hard work.

I liked the plastic dairy cow they had at the show. I don't know where she came from, but one thing was sure, she was good-natured and did not kick! It was a great idea in terms of education. I saw a young lad struggling to milk the cow so I showed him how to squeeze the milk – or water – from her. It might be a good idea to take her to the schools if she is being over-wintered in Uist!

I was not happy to see NFU Scotland at the show. I know mutual insurance is a good thing, but NFU trying to sign up members is a bit insulting. I hope crofters understand that it was the NFU that demanded a three-tier region payment while the Scottish Crofting Federation (SCF) fought hard for a two-tier payment. *Comhairle nan Eilean Siar*[8] also backed a two-tier system, but the big boys in the NFU persuaded Richard Lochhead to make it three, and so we are in a mess with many still waiting for their payments. Do not forget that the calf subsidy you get would not be there either if it were not for the SCF and Quality Meat Scotland.

I was shocked when the EU referendum results were revealed. It is a nightmare, absolutely unbelievable. Unity has never been so important and necessary, and here we are breaking links. We hear our government arguing with agriculturalists and fishermen that unpopular rules are EU directives so let us see how smart they are now that they can't blame them.

Paible School children and staff had a busy end-of-term session. The local mod was in Lochdar School, a concert to mark its closure, and the annual prize giving day, and *Feis Tir an Eorna*. What a talented bunch they all are, real stars, well behaved and always welcoming us oldies with a charming smile. I love the way the Head Teacher asks the children to put their hands on their knees so that they sit still, listen and concentrate.

At the prize giving, nobody is left out. When I was in school, three books were presented in each class, first, second, and third, and the academically bright got them every time! It was sad for the others with perhaps more common sense and practical skills which would be useful in later life.

I have attended many happy occasions in that school, and like to watch all the children and try to work out who their parents are. Thirty years ago I used to recognise most of them, but now I struggle. Sometimes I see some resembling their grandparents, but there are many families that have moved to the island, which is good because it introduces much-needed new blood.

I listened to a programme about Amazon's plan to use drones to deliver goods to your door. What if they took the wrong route? They might bring down an aeroplane. I wonder if anyone remembers the cartoon of Angus Og, something about a crofter dreaming of the peats flying home from the moor and forming a beautiful stack! Maybe it will soon be reality. The drones could be handy too when an animal escapes, opening the gate and driving the animal back inside while you sit in your armchair pressing buttons.

September

It seems like yesterday that I was looking forward to seeing friends and family here on holiday. Now they have gone back, the place feels emptier. But nowadays we can contact our loved ones by telephone and modern technology that's beyond my ability! Of course, I can still look forward to Jessie, David and family coming back soon. They are almost resident here now.

In the 1950s and 1960s, most people here didn't have a telephone so the only way to keep in touch was by letter. And it's still nice to receive a letter. I have often been teased about a letter I sent to my sisters, Flora and Agnes, when they spent a summer holiday helping our grandmother in Mull. I had only been in school for a few years. Granny Mull hosted tourists in the farmhouse so there were many dishes to be washed. She used to get ulcers on her legs so my letter read: 'Dear Flora and Agnes, I hope you are well how is Granny legs the cat had kittens!' At least I made the effort.

Alexander goes to the machair every day before school to chase the geese away. It's never ending and this is the last year we will get crop protection funding. It is now illegal to put goose in the food chain. People were enjoying eating goose burgers and sausages, but this government would rather waste good food. Our young crofters will need to start strong lobbying to get the scheme going again. When something is working well, why is it then stopped?

The primary children are now in the new school and I hope they will have many happy days at *Sgoil Uibhist a Tuath*[9]. Meanwhile, the demolition squad is busy knocking down what is left of the old school. When I drive past, I wish I had blinkers to stop me seeing what's happening.

Two of my old hens died recently. They were over fat from eating too many cooked potatoes. Onslo – the cockerel I got from Cath Major – is now too old and has lost all interest in the hens so I haven't had chicks for two years and although hard work, I do miss them. It doesn't seem that long ago that I showed Alexander and Archie MacLellan the day old chicks in the chicken house. They were desperate to cuddle them and were so excited when I let them hold one each in their chubby hands and they held them close to their cheeks. Children miss out on these simple joys these days.

October

Harvest has been such a struggle. Crops to be made into silage bales are always better secured when dry, especially the cereals, but it hasn't been possible. If it didn't rain in the morning, it rained in the afternoon. Combining was even more difficult because the corn must be completely ripe and dry. This year that was impossible and lots of the seed fell on the ground because of the strong winds and driving rain. It is very clear that the weather is changing. We always had good and bad in the past, but the wet and wind were not so constant, and we had more sunshine.

When I was at school, before anyone here had binders, the only implement available was a scythe. I heard the men say that Calum MacDougall could sharpen a scythe better than anyone else. Unlike with the binder, the wind direction did not matter and a man and scythe were very flexible.

After school we would go down to the machair, walking of course, and we could carry a wee tin pail containing clear cold well water mixed with half a cup of oatmeal to quench our thirst. The adults would have been hand-binding sheaves all day so it was up to us to make the *suidheachain* or stooks. One of the advantages of the old method was being able to cut the corn before all the seed was completely ripe because the seed ripened fully on the sheaves in the field. Also, when making *toitean* or ricks – small ricks of about 120 sheaves – bone-dry sheaves went at the bottom, and damp ones on the top. If the ricks were built properly, in a couple of weeks they would all be thoroughly dry and ready to take home by horse and cart where they would be made into big stacks, usually in the yard behind the house.

We loved standing in the cart while the adults threw the sheaves to us. We would skilfully shape and build the load, the seed towards the centre and the cut edge of the stems to the outside, and it was layered to bond the sheaves in case any slipped out. We called that *grabhadh*. The top row would act as a roof to secure the load with two ropes thrown from the rear of the cart, and tightened to make sure it was in the right position. Then we would lie down on top of the load, and hold on to the rope in case we fell. This was the bit we loved!

Most carthorses were strong, steady and calm. At the stack yard, if the weather was adverse, the load would be tipped out, and because it was so well built, it would slip out and the rain wouldn't penetrate. We often threw an old iron rim of a cartwheel over it to stop the sheaves blowing away. With big *cruachan* or stacks, dampness was not a huge problem as long as the foundation was bone dry. At a certain height, the damp sheaves were put on the outside row, and as long as the central rows were higher, firm, sloping downwards, closed up with sheaves of marram grass and tied securely, that stack would be perfect when fed to the livestock in the winter.

There were more people around then, and we laughed and chatted happily. The only things I hated about harvest time was binding barley because the ears stuck to our clothes and body, and the thistles hurt our hands. And the midges could be a plague!

Our problem now is that we are an ageing population. There are fewer crofters and most have more than one croft and more cattle to feed. Without machinery, the harvest can't be done. In my day you could always find someone to help. Cottars[10] without land would help and got corn to feed their only house cow in exchange.

171

November

Recently the seed dryer has been working overtime with Alexander helping his dad. I was pleased to learn that he wears a mask to protect his lungs from the dust.

We had a Welsh film crew here during September. Alexander was in his element, acting the perfect host, and showing the filmmakers how to gut and prepare fish for dinner, something they had not done before, so Alexander was quite proud of himself. He is fond of fishing and hunting, and I am glad he is keen to do the preparation for cooking as well. If his great granny were alive, she would be very impressed.

Alexander and his parents were in Germany recently. His dad was judging at a Highland Cattle Show so it was a good opportunity for a holiday. Our farming friends over there live in a beautifully restored, seventeenth century farmhouse. Alexander brought me an ornament of a typical German wooden house set on a base with wooden trees in the garden.

'Granny, look at this,' he said, as he removed the house from the base to reveal a tiny metal dish. He explained there was 'stuff' you put in the dish which, if lit, would make smoke that rose out of the chimney. Then his face dropped and he said, 'Dad didn't want to get it in case you set your house on fire.' I'll need to find a tiny candle so that we can see smoke come out of the chimney.

Our cattle have now returned to Vallay for the winter. We move them over gradually with the first lot crossing the strand in mid-September. They were already at Dusary so it made sense to walk them across the Committee Road. On the day, the tide was fully out at 11.30 am. It is usually shallow enough to cross, about three hours before low tide, but there's still a lot to consider before take-off, especially the best time to go to avoid the traffic. After careful consideration, we determined the best time was 7.30 am. I was leading the cattle in my Landrover, with Angus, John and Norman at the rear. Ruth, forever helpful, stayed at the Dusary end of the road and advised any driver heading across hat it would be wiser, and certainly quicker, to go around the west side because there was no way they could overtake the cattle. None of the cattle had walked the road before. Although used to tractors and jeeps in Vallay during the winter, they were very reluctant to pass parked cars, especially cars with sidelights. If they stopped walking, I called them by name, and they carried on. I could see the beasts were frightened by the cars so when vehicles came towards me, I politely asked the drivers to stay in the passing place as close to the verge as possible and turn off their lights and engine. Someone suggested we should shift the cattle at 5 am before the traffic starts moving. I doubt that person realises it would be pitch dark and the tide would still be in. Eventually we arrived at the Sollas end of the Committee Road at 8.30 am, just as the school mini-buses were turning in, and the cattle crossed the main road onto the strand. The tide was perfect and it was not raining for a change. Soon the cattle were over and up to their knees in luscious grass, with plenty of water and shelter. I'm often asked how we get the cattle to Vallay. Some tourists ask do we take them in a boat! Anyway, I thought I should explain how it is done, and thank all the drivers who were co-operative and patient. I hope none of you were late for work!

We had a very special happy celebration in Kyles on Friday 21st October when twin girls, Hannah and Margaret MacDougall, turned twenty-one. Family and friends enjoyed a lovely evening in Carinish Hall where the Stepping Stone restaurant served its usual first class food and Calum Iain performed lovely music. The girls' granny, Dolly, had her birthday the next day. Congratulations to the trio!

December

On Radio nan Gaidheal, Dr Kate Dawson spoke about Lyme Disease. A few days later, local vet, Graham Charlesworth, spoke about his research. There have been 168 cases in Uist over the last four years. We agree ticks are being spread by deer and there are far too many deer, especially in North Uist. When we were young, we spent a lot of time on the moors helping with peat work. Doctors today advise people to cover up to guard against ticks, but we never wore long sleeves or trousers and we were often barefoot. We were not afraid of ticks and honestly, I had never heard of Lyme Disease. When we were out on the moor, we never saw deer. There are so many now that we see them everywhere, including in gardens and on the machair. I must have been thirty before I saw a deer on Uist. Now though, when people are leaving early in the morning to catch a ferry, they're told, 'Be careful you don't hit a deer.'

In the early part of the twentieth century, large crofting families needed plenty meat to survive and they were prepared to take risks. Without electricity, venison had to be eaten within a few days so was shared among the townships. The crofters believed that the deer belonged to the landlord, but in the mid 1990s, the chairman of the Deer Commission, Patrick Gordon Duff Pennington, informed us that the deer do not belong to anybody, but the estates own the shooting rights. However, crofters have shooting rights if the deer are grazing on their improved ground. He was quoted as saying, 'It's time for the North Uist crofters to stop shooing and start shooting.'

Alexander was recently very excited to bring me a number of cuddies. He hadn't seen them before. He and his dad fished with a rod and winkle bait off Griminish Pier. When we were young, we used to fish for them by rowing a small wooden boat to Sgeir a' Chotain near the Kyles sea pool. We got dozens of them.

'Alexander, I'm going to enjoy these and you've gutted them already,' I said.

I told him how in the early 1960s, his uncle Gus was here the year he and Agnes got engaged. Gus went with his rod to fish for *saithe*[11], hoping for a sea trout but came back after a couple of hours and said he had got nothing but tiny white fish about the size of sardines which he'd thrown back in. Agnes, although disappointed, had a good giggle. On a visit to a friend in Bayhead, Flora Ann MacDougall, she told her about Gus throwing the cuddies back.

'Oh dear me,' exclaimed Flora Ann, almost in tears. 'When I was young, sometimes the cuddies were our main food.'

And I know folk from Paiblesgarry used to catch cuddies with a hand net using cooked potatoes as bait. I have never seen that being done before, but they caught hundreds.

This is the last article for this year. I wish all my readers a blessed Christmas and New Year.

2017

February

It's hard to believe that Christmas is over for another year. We had a tasty, healthy dinner of wild ducks and pheasants served with our own vegetables from the garden. Alexander was kept busy preparing and plucking the ducks except that he and his dad ran out of gas while singeing them in the shed. Ever resourceful, Alexander came running in with the orders!

'Granny, you have to finish them,' he announced!

Luckily I have a small primus so that did the job.

Alexander said afterwards, 'Granny, next time I get a goose or a duck, I will gut it if you show me what to do.'

I will make sure he sticks to his word!

Recently we had a visit from two Italian farmers, brothers who live in the far north of Italy, near the Austrian and German borders. They showed us photographs of their Highland Cattle up in the mountains, covered in snow. I noticed that their cows wear bells around their necks. I do not know how ours would react if we put on bells. They would probably stampede! The brothers graze their cattle in the mountains all summer and all the farms share the same common ground, just like crofting townships have common grazings. Cattle from different farms are grazed together and a herdsman is paid to look after them. Cows run with the bull both before and after they spend the summer on the mountains. This was the only family in their region to keep Highland Cattle. The others have Brown Swiss, a dual-purpose breed, dairy and beef. Their second last bull was bought in Germany and was actually bred here at Ardbhan.

I was hoping to buy some salt herring this winter, but I could not find any in the shops. At the beginning of January I was listening to Coinneach MacIomhair's programme on BBC Radio nan Gaidheal. He was chatting to Peggy Margaret – MacKillop, from Lewis, who was that day celebrating her hundredth birthday. She was wonderful and you could definitely tell that Coinneach was enjoying the conversation with her. Who would not? She alked about the huge quantities of herring that used to be landed t Stornoway when she was growing up. Coinneach told her that e heard that herring had been landed on North Uist that same veek. So after a quick phone call to Kallin Shellfish and a trip o Grimsay, I am now enjoying some very tasty salt herring. Coinneach MacIomhair must be an absolute mine of knowledge - he should compete on some of these TV quiz shows!

My collie dog, Rocky, is now about five years old and this vinter he has been glued to the television in the evenings. He oves to watch Donald 'Sweeny' MacSween from Ness in Lewis on the 'An Lot' programme on BBC ALBA. Rocky gets excited is soon as the programme begins, especially when he sees or hears the sheep and hens. If my phone rings while 'An Lot' is on – and I put the television to mute while I answer the call - Rocky starts whimpering and stands on his hind legs until I but the sound back on! I like the way Sweeny handles his animals. His genuine interest and love for the sheep, cattle, pigs and hens is obvious.

March

In 1996 when our German farmer friend Axel brought two heifers born in 1994 here from Douglas and Angus Estates, who would believe that one would still be here in 2017? Until 1998, all Highland Cattle had a tattoo in one ear. One of Axel's heifers had HOHF3 and the other heifer had HOHF4. HOHF3 is the survivor, her pedigree name is Seonaid Ruadh. HOHF4 was called Peigi. Although HOHF3 is 23 years old, she is still in beautiful condition, her feet are perfect, and she looks in-calf. Her granddaughter, Seonaid, born in 2003, has never had a problem either, until her male calf was born in 2016. She got mastitis and the calf wasn't getting enough milk, so I gave him a bottle as a supplement. They were over in Vallay, and as soon as he saw a jeep, he would come running for his bottle. I stopped giving him milk mid-June as he seemed to be doing fine with his mother. A few weeks ago he was brought home here and put in a pen in the shed along with two other calves.

I went out to see him and called Moicidh Beag and he pushed his way through the crowd and came running over to me. He started sucking my fingers even though I hadn't had any close contact with him since June. I was amazed he still remembered me. So you might guess what I did .... I went inside and warmed up a litre of milk and took a bottle out to him! He sucked my fingers first and then grabbed the teat, so he enjoys his milk every day now. He is small, but his back is well covered and it's great to see him happy.

We have been hearing about the Crofting Commission in the news almost daily for the last year. It's been very difficult to grasp what has been going on in Lewis. However, it will soon be time for us to vote for a new commissioner for the Western Isles. Considering the geography of the islands, it doesn't make sense to have only one commissioner as the time spent travelling could be better used. It has been a while since we had a commissioner from Uist but this year we have an opportunity to change the situation. Our candidates are Alasdair MacEachen from Benbecula and Iain MacIver from Lewis. Alasdair is well known throughout the islands and is a very able and active crofter. He has done a lot of work for crofting since the Scottish Crofters Union was set up in 1985. Through his involvement and experience, he would be the ideal candidate.

I was privileged to be invited to the official opening of Sgoil Uibhist a Tuath last month. I felt a bit hypocritical because I have felt sad at the old school, which opened in 1963, being demolished. It was emotional to see Rev John and Mrs Margaret Smith unveiling the plaque. Rev Smith opened the old school in September 1963. Himself and Margaret are a special couple much loved by all in Uist. During the opening ceremony, the children sang six Gaidhlig songs and everyone enjoyed their performance. After the formalities, there was tea and coffee and the most beautiful cakes baked by young Eilidh Boyle. I think the last time I saw cakes like that was when I worked in Glasgow in the late 50s and used to window shop at the Fairy Dell Bakery and Tea Toom in Partick.

April

Birds are singing as they start to prepare their nests, daffodils are in full bloom, and I have seen primroses over on Vallay. Early born lambs are getting a good start to life when the weather is pleasant.

Greylag geese are pairing off. When this year's goslings grow up, there will be more than ever. Fields everywhere are polluted with their dirt. Barnacle geese do not breed here but they will still be around for another month, long enough to destroy the early bite. Thousands now spend the winter on North Uist, causing a lot of damage, particularly on the most productive fields. It seems they love short, green grass. We never see them in thick, long, rank grass.

Our Uist candidate was unsuccessful in the Crofting Commission election, losing by ten votes. I still think we need one commissioner for Lewis and Harris and another for Uist and Barra. It makes no sense for one person to travel that distance.

When I was a child, communion weekend was held in three churches at the end of July. In the 1940s, we did not have causeways, and motorcars were still quite rare. Two visiting ministers would come to North Uist for the weekend. Our own resident minister came from Scalpay so many friends came across in their fishing boats. The manse would be packed and Kyles and other villages would be teeming with people staying over, mainly from Grimsay and Harris. I well remember when Ruaraidh Uilleam and his son Lachlan MacDonald (Cnoc Ard) stayed with us. Ruaraidh was an able Gaidhlig psalm presenter with a strong, musical voice.

Young people may find it hard to imagine the preparation before that weekend. We didn't have electricity or running water. New potatoes would just be ready and sheep would be killed. Trout was served, if caught, and hens and young cockerels. Back then, you could not drive to a shop to buy ready-made food! Visiting fishermen would bring a great variety of fish, much of it salted, and lobster. And young rabbits.Thursday was called Latha na Traisg, which means the fast day, and only necessary work was done. There would be a service at 12 noon and 6 pm. Friday was the question day, Latha Cheis, when five members would give a short sermon on a verse chosen by the minister. This was the longest service, lasting about three hours, and to be honest it was not a popular day! The Saturday service at noon was as usual, but many of the women stayed at home to prepare food for Sunday and milk the cows and the feed calves. Sunday morning was usually a three-hour service, and another two hours in the evening. All the services were conducted in Gaidhlig. We enjoyed this special weekend, especially on the Sabbath when we got dressed in our best clothes, our hair beautifully styled with ribbons and, of course, women and girls wore hats. We only got sweets once a month, so the weekend visitors brought some and that was a real treat. All weekend we would eat like royalty! I believe the services were too long and we would sometimes be too tired to concentrate. But we enjoyed it.

May

A lovely sight greeted me when I walked into the big shed two weeks ago. There was my grand-nephew, ten-year-old James, sitting against a big round bale of hay surrounded by twenty yearling heifers. James was sat with a bundle of hay on his lap carefully feeding handfuls to the more tame heifers. I am sure they were in their element, being well looked after by their new friend, and probably thinking their lives could not get much better.

Yellow Myra, the friendliest calf in the bunch, certainly made sure she got her fair share of the feed, but because she has three-inch long horns, I warned James to not let her come too close to his face. I was over in the shed topping up their water containers and of course as soon as I turned on the tap connected to the hose, James' younger brother, Alister, wanted to hold it. He thought it was magic squirting water everywhere. I was surprised that he didn't overfill any of the five tubs! While all this was going on, eleven-year-old William was standing by the bull pen where three young bull calves were curiously putting their noses up to his hands. James was sad to return to Inverness and leave his new friends, although I doubt his mother would have welcomed Yellow Myra or the black bull, Muran, into her sitting room!

While the visitors were having a great time in the shed, Alexander was out spreading seaweed with his father. Unsurprisingly, young cattle are not a novelty to him!

Early one morning at the end of March, Black Sobhrach gave birth to a black female calf. Later in the day, myself and Ruth did our usual rounds while the tide was out in the afternoon. We saw Sobhrach and her calf grazing near the top of one of the highest dunes, a common sight on Vallay. After we had checked every secluded spot, we were ready to go home, and driving along the track we could see that the calf was right at the edge, and over that edge was an almost perpendicular sandy drop. While we were trying to decide what to do, Ruth saw the calf fall over and roll down the slope. Sobhrach saw what happened and stood at the edge looking down at her calf. If she dared follow it down, the calf could be in grave danger so while I turned the car round and drove to the nearest place, Ruth climbed up the grassy slope to the right of the cow to try to drive her back. I took out a feedbag and called Sobhrach. She rushed towards me and followed me on the track that led to the bottom front of the sand dune. I gave her some food, and then fixed my eyes and mind on the calf which was not even a day old and had climbed back up to the highest ledge, just below where she had fallen. But couldn't get any higher. Brave Ruth grasped the situation and thinking only of the animals, scrambled over the edge, grabbed the calf, held it across her body and slithered down the sandy slope to where myself and the cow were waiting at the bottom. Sobhrach had just finished eating and was happily reunited with her calf. Despite the danger, it was funny to see a woman and a calf sliding down the slope like a TV outdoor challenge! We both chased the cow and calf well away from the area and back towards the other cows. Never a dull moment when working with animals!

July

Last month we saw the second annual Tractor Rally on North Uist. All money raised will be used to support sufferers of Motor Neurone Disease and fund research to find a cure. It was an overwhelming success and very emotional. I met and spoke to so many people who have lost their near and dear to this horrible disease and yet they were so brave and serene in their grief. I cannot find the words to sufficiently praise the committee and volunteers of North Uist Agricultural Society who organised the rally. It was a mighty job and the money literally poured in from our very generous community. I had watched the tractors setting off in the morning from Hosta, the older and slower tractors leaving first. Several did not have cabs so the drivers were well clad and prepared for the torrential rain. I joked with Rev Lachie MacDonald and Rev Calum Smith that they had not prayed enough for a dry day! I love the photograph of Lachie in the June edition of Am Paipear. He is smiling gleefully like a little boy sitting on his new toy tractor on Christmas morning! After the tractors had set off, I went home to feed the small animals but kept looking across from my window towards Horisary. Heading south, one special tractor caught my eye. It was towing a trailer upon which sat an old grey Fergie. It brought tears to my eyes. It was as though the stronger tractor was giving the weaker one a piggy back and that is what the Tractor Rally is all about. Surely the energy and love shown that day will help bring about a breakthrough in finding a cure.

On 21st June, all the cattle came across from Vallay. This is the hardest and busiest day on the cattle calendar. I stay well clear of it all now and just do the preparation paperwork. We work with five bulls so it takes a lot of thinking and writing, making sure the pedigree is correct, and that none of the bulls get to their own progeny.

Two of the older cows were brought to Kyles to get medicine. I emptied the water trough so it would fill with fresh water because there is no stream in the small field where they were going. These cows had never been to Kyles before so I had to entice them to the water bath. They followed me and must have been thirsty but when they saw the bath, they stared at it as if they had seen a wild monster! Even with me splashing the water with my hand, they turned around and ran away. Eventually, they will understand it is water and will drink!

I have not seen a rat for years, but last week I saw one outside the henhouse. It had dug a tunnel right underneath. Having had experience of sheds being overrun by rats, I set a trap with cheese for bait and caught this one. With a fork in the trap handle, I carried the trap towards the sea. But the trapdoor, which opens under the weight of the rat as it enters, swung open and shut so I was worried the rat could escape. There was a lady with her terrier on the beach and I was confident that the dog would catch the rat before it got away. I asked the lady, and she abruptly told me no! She asked if I was going to let the rat go on the beach, and I replied that if I intended to let it go, I would not have gone to the bother of trapping it in the first place! I think she thought me barbaric for drowning a rat!

Rats and agriculture are a very bad combination. They can destroy silage bales and feed bags, pollute dry fodder in sheds, cause disease and kill chickens. I better not write any more about rats. If anyone is interested, ask me what happened here in the early 1940s when we had a plague of them.

Now that the government is not giving any money towards crop protection, this coming harvest could be a disaster. We read of crop failure in hot countries because of drought, but here we could lose most of our crops unless the geese are culled. This is extremely serious. What will be left to feed livestock over the winter? There is almost one million being spent to protect three pure Scottish wildcats. Sea eagles and ravens are also protected by law. It is time we rose up to protest or our crops will end up as a wilderness and the predictions of the Brahan Seer[1] will come true.

Here, three couples are getting married. My own grandson, Fraser, and Carianne will get married in Inverness. It will be good to meet together at the happy occasion. God bless them all.

September

I have never seen the ground so wet in August. After the heavy rain, the river Horisary was flowing across the road and there were pools of water on the machair. Some crofters have had cows trapped behind streams and ditches, yet astonishingly they can find their way to safety.

Environment schemes such as restrictive grazing are destroying the wild flowers and the overgrown grass is less nutritious. Policy makers set a date for harvesting corn or hay, but should we not harvest when the crop is ready and the weather is suitable? Woe betide any crofter who cuts their crops on 31st August after signing up to not cut before 1st September. They will be penalised by losing their payments.

Am Paipear has a lovely photo of schoolboy, Archie MacLellan, at the North Uist Agricultural Show at Hosta with his champion ram, judged and chosen by Mel Irvine, everybody's favourite TV star in the BBC series This Farming Life. Beneath was a photograph of Archie receiving the Ivan MacDonald Memorial Trophy from Annie MacDonald, the late Ivan's mother. Archie looks so happy. He is a tonic for my age group. It is obvious he loves and knows his animals and I wish him every success. Ruairidh 'Glebe' MacDonald won the cattle section with his champion Limousin heifer calf. The future of crofting is safe in these young hands!

Onslo, my old white cockerel, has survived another winter, in fact he looks better than he did two years ago. Alexander found a hen sitting on eleven eggs. We decided to leave her as I did not think the eggs would be fertile anyway. Despite the constant rain, she made a secluded nest on the ground, sheltered from the wind and rain under an old tractor cab. Just before the end of August, we were surprised when two little chickens appeared, although the other nine eggs were rotten. I have never handled such a wild broody hen. I go into the stable to feed her and she jumps at me! My hands are now covered in bruises but at least no cat would dare go near her. After three years without chicks, it is nice to have some again.

Last month I was asked to give a talk at the opening of the Toradh[2] festival in Benbecula. I spoke about the food we lived on when I was younger and how our diet varied at different times of the year, depending on what was available. Shops were not so well stocked back then and in the days before fridges and freezers it was difficult to store meat. While everything might have been less convenient, our diet was almost all homegrown and healthy.

Visitors, friends and family are now returning to their mainland homes and we miss them. Roads are not so busy with autumn well and truly with us.

October

I think Cupid has been working overtime this year with several Uist couples getting married. I wish all the new couples joy. I attended two weddings this year. My grandson, Fraser, and Carianne's wedding, earlier this summer, was the most important. Carianne was stunning, quite a contrast to the way I usually see her in a dirty pair of jeans and wellingtons out feeding young bulls or driving the ATV! Fraser was very handsome in his kilt. It was lovely to be there with families, relatives and friends. Alexander was smart in his kilt. He enjoyed being an usher and took the job very seriously!

Last month I was at the wedding of Ross Morrison and Ashley Alcorn in Glasgow, and took a ten-day break in Renfrew with my sister Flora. The bridal party looked wonderful in their finery and again there were many guests from Uist. No one looked more elegant than Granny Katie in a beautiful crinkle dress with a fine lace covering. I danced a lot at both weddings, and at Ross's wedding I had a special request with Uncle John, Old Time Waltz to a Gaelic tune!

I enjoyed meeting old friends and cousins in Renfrew. I went to Ayr to visit my late sister Agnes's family. It was good to see Gus and all their young grandchildren. I miss Agnes so much.

Some of you might remember the story of the black Highland cow, Margaret Fay Shaw, who gave birth to a deformed calf that could not stand up until it was six weeks old. The cow had never been handled but when I realised the calf could not stand, she allowed me to handle her teats while my friend, Ruth, supported the calf by holding her hand under its brisket.

Sometimes the cow is a good distance from the calf but I thought of a plan. I had an old, brown shopping bag with a zip. I put some feed in the bag and she followed me down the valley. It was quite comical. I would put the bag on the ground to open the zip and eventually she pushed the zip open herself. While she ate, I would help the calf up onto its feet and it sucked perfectly.

Earlier this year when we were moving cows, Margaret was put with the wrong bull. I wanted her with Big Boy. It is more or less impossible to take a single cow out of a group of about twenty and move her a few fields away. Then I remembered the previous plan – the old brown shopping bag! Maybe I was being too optimistic. Would Margaret remember the bag? I decided it was worth a try. Ruth came with me and as usual the cows were right at the back of the field. I often speak to Margaret and give her a wee scratch, and the rest carried on grazing and were not a bit suspicious. I put the bag to her nose and I knew immediately that she remembered. Ruth walked behind her and she followed me like a pet lamb right to the road side, out the gate and down the road to join the others with Big Boy.

So for those people who think that animals are stupid, this proves they have good memories, even photographic memories! Margaret will be seventeen in December.

Most of the harvest work is done now and I hear there was a good crop on most crofts. Potatoes are also good this year and will have to be lifted soon and then the cattle can get to the machairs. Summer was not very warm and I didn't see cows standing in the shallow water trying to cool off.

Some of you might remember Terry Williams who had a stand at North Uist Agricultural Show a few years ago and spoke with crofters who used to walk their cattle to the markets. She has now written a book entitled Walking with Cattle – In search of the Last Drovers of Uist. She sent me a copy a few weeks ago. There are some lovely old photos along with some interesting stories.

Alexander's two chickens are thriving.

Now the nights are getting longer and I am already dreading the winter gales. I hope we do not get fierce storms. At the end of October the clocks will go back an hour. I wish they would leave it like that all year round instead of changing it twice a year.

November

I didn't mention geese last month but I cannot ignore them. There is an open meeting organised through the Scottish Crofting Federation. I know in the past there have been meetings and no action. Geese are counted and SNH tells us that large numbers have been shot. However, they forget there are probably more goslings being hatched than were shot and so the population grows again!

Recently I saw more barnacle geese than I've ever seen in one day on the machair at Kyles. In one twelve-acre field where silage had been cut there were some 1500. Greylag geese had been grazing constantly throughout August and September so the field never got a rest, and since the barnacle geese arrived in October, the grass has not had a chance to grow for young cattle. There are sheep in the field now because the grass is too short for cattle. Heading west, three fields along, there was a similar number of barnacle geese, and the same number on the other side of the dividing fence. While I was on the machair I heard gunshots further west and the second bunch flew up, circled, and landed beside the third group a hundred metres away. There are about 8000 barnacle geese on Uist, predominantly on North Uist.

I have a copy of a letter that Roseanna Cunningham, Cabinet Secretary for Environment, Climate Change and Land Reform, sent to our MSP this summer. She has decided that because the adaptive management scheme did so well in reducing numbers, we should be happy, and it is up to crofters to manage. Each year the goose situation gets more sickening.

I had my three young nephews from Inverness here for a week of school holidays. What a help children are when they enjoy their jobs! It did not take them long to gather the electric string and plastic pegs that controlled the young bulls during the summer, and they cleaned the hen house well using a spade and wheelbarrow. They have a vegetable allotment in Inverness and the highlight of their 'work experience' on North Uist was when we lifted the machair potatoes. Alexander and the boys enjoyed that task especially at first while it was a novelty. After the job was finished they ended up behind my house jumping on the trampoline. I enjoyed seeing them happy and at the same time being a good help. I probably won't see them again until Easter when James will enjoy feeding the heifer calves in the shed and Alister can help me fill the water troughs. No doubt we can also find a job or two for William!

My collie dog, Rocky, had a nasty accident about two weeks ago. He cut his shoulder and hip on some sharp wire, but thanks to the vet, David, he is now back to normal. I have also been nursing a white hen for weeks. Something was wrong with her legs and so I've been putting her outside in her own pen. In the early evening, I open the door and while I am putting the other hens inside, Rocky goes into the pen and takes the white hen out as if she's a lamb.

One more reminder to you young crofters – and the not so young! Over the last few months a consultation on the future of crofting law has been ongoing and will continue until 20th November. I have a hard copy of the consultation document and it can also be accessed online. It is not simple, but it's very important.

December

I opened the small hen house one morning a few weeks ago and found a blackbird flying around inside. I tried to get the blackbird out the door but it would not leave. Some rats are still scratching around and one continues to make a tunnel in the sandy floor. Having refused to leave the hen house, the blackbird hid in a corner and, as I watched, it reversed over the rat hole and suddenly disappeared. The tunnel must have been perpendicular for the blackbird to vanish as it did. Then the blackbird started squawking so I bent down under the perches, put my hand into the tunnel and, although I knew where it was, could not reach the bird. What a noise it was making! I rushed out, grabbed a spade and began digging into the sand. I reached into the hole again and this time managed to retrieve the blackbird. It was so frightened! There were two cats prowling nearby and I did not want to let the blackbird go in front of them. I walked away from the hen house, but Cissy, the mother cat, followed me as she usually does and the more I tried to chase her away, the closer she came! I was reassured the blackbird was not injured as it fairly struggled. I lifted it high, facing the wind, as the planes do, and away it flew. It was great to see the poor wee creature fly to safety.

I don't think Cissy knew I had a bird in my hand after all.

Alexander's two chickens are definitely pullets, which is a relief as I was dreading that one would be a cockerel. Rocky hangs around with them for hours. Last week we had some dry weather and I left the door open while I went to get some peats. When I came back , Rocky was having a great time. He had taken the mother hen and two chicks from the yard into the kitchen, and was standing in the middle door determined they would not get out!

Later that evening when I let Rocky out for his last walk, I opened the door and he refused to go. He just stood there as if somebody was at the door. I put my head outside and who was there but the old dun cow TeePee and her calf standing beside the barrow of peats, not a care in the world, busy chewing her cud. It was as if to say, 'If the chickens are allowed in, why am I not?' What next? Rocky would be a star in a nativity play, bringing all the birds and animals into the stable.

There was a good turn out at the goose meeting on 21st November and I think Eileen Stuart (SNH) got quite an eye opener. We had a good discussion and she was told that crofting would cease if

SNH ignored our problems for another year. I liked the point Alasdair MacDonald raised. He asked the SNH staff if they ever think of the psychological impact of the geese on crofters. I had never thought of that but it was a very wise question because it does affect us all year round and all we hear from SNH is managing, counting, scaring, monitoring, numbers increasing and decreasing, and so it goes on and on. I hope and believe that they now know that we will not accept any more talk and that action is their only option.

Myself and Sarah attended the cattle and sheep sale at Lochboisdale on 20th November. Trade was very good. We have some wonderful, generous people on our islands, and who more generous than John Iagan Gillies, a fisherman and crofter from Barra. He sold a growthy[3] young Aberdeen Angus bullock for £720 and donated with such sincerity the proceeds to the 1 Million Miles for Ellie fund.

Our country is in turmoil and we are all scunnered with Brexit. I do not understand our Prime Minister, Theresa May. She voted remain in the EU referendum, and then became Prime Minister and put her passion into something she did not believe in. She had a couple of opportunities to make a u-turn but did not. Maybe the Irish will save the day?

Angus on his tractors, aged 4 and 21.

Ena and Lasair
Chlach Chanaidh.

Angus selling
Pibroch of
Achnacloich.

Ena and The Rhumach.

The famous Kirkibost of Ardbhan.

Angus on the Ranch.

1995: Ellie and Fraser.

2010: Alexander and the
black and white puddings.

Belle the pet hen.

Rocky and Cissy.

2006: Ena receives her MBE.

Ena with her sisters.

Ellie's graduation with Alexander.

Alexander, Angus, Ellie, Fraser, Michelle, Sarah.

Ena is 80.
Angus, Sarah, Fraser,
Alexander, Carianne,
Ena, Michelle.

2018

April

There is a great deal of ignorance about muirburning. Throughout my childhood, every February and March, there was a lot of burning and we never questioned why it was happening apart from a rough idea that it was to stop the heather growing too tall. In the 1950s and 1960s there were good crofting grants and in the early 1960s the Kyles crofters fenced 150 acres of hill common grazing. That area on Marrival had not been burnt for a long time, the grass was choked by the tall heather so cattle did not thrive there. in the 1970s we decided to put cattle there and walked them all the way on the hottest day of summer! Eventually we arrived, but the animals looked miserable so after ten days we took them home. I knew it had been a rich grazing area so after some research with older crofters, we decided to burn the whole area the following March. I contacted North Uist Estate and found some volunteers. With the tree plantation, the burning had to be carried out carefully in an easterly wind which was not so common back then. I wasn't there, but as I watched the smoke rising, the wind suddenly changed to the west! Panicking, I wondered who would stop the fire spreading to the trees. I phoned North Uist Estate and someone sent for help. At the end of the day, the whole area had been burnt and not a spark had reached the trees. Phew!

Next summer, lush green grass grew and in July we sent the cattle for another trial. It was a huge success. Cattle were sent there every summer until 1997 when the fences became irreparable.

I strongly believe that if there was more muirburning the deer would not come as close to the in-bye land. It would be good to have sheep there too but without a full time shepherd they would have little chance of survival with the ever-increasing number of eagles.

Theresa May now has a Brexit date for next March. It is becoming more and more scary. Surely there is some strong person who could persuade the UK to remain as it is. We need another Churchill! We need politicians who speak the truth.

202

What is happening with CalMac? Disruptions are more common every year and the public is fed up with it. The winter weather has been calm yet cancellations are more frequent. Thinking back, the Lochmore and Lochear ferries of the 1950s didn't even have stabilisers and they managed to reach their destinations.

Alexander is very busy. He is becoming an expert with the wheelbarrow and shovel and helps Carianne keep the young bulls and heifers in the shed clean and tidy. It's healthy for young ones to work with animals. It develops their muscles.

May

Back in the 1980s, I used to give a few primary school children a lift to Paible School. One young lad talked about his pet lambs. I asked him how this year's lamb was getting on and he said that every year his pet lamb goes to the hill with all the others and never comes back. Later I realised that it wasn't the hill near his home but HILL at the market! Maybe thinking the lamb was out on the hill made it easier for him to part with it.

This little boy is now one of the most popular Gaelic presenters on radio and TV. His Gaelic is fluent and correct and he speaks with a perfect North Uist accent. Perhaps you have guessed who I am writing about? It is of course Darren Laing, an excellent ambassador for the Gaelic language and North Uist.

We have a six-year-old small brown Highland cow who is as tame as a pet lamb. When she was a yearling, she took a nasty condition called coccidiosis which was cured with Vecoxan. However, she was still not thriving. Her coat was dry and she seemed lethargic so we asked the vet to take a blood sample. She had a low level of the trace element selenium. Once this was corrected, she was a new animal. She had her third calf last month, but instead of suckling, the calf licked her mother's coat, even though the cow's teat and udders were in perfect shape. We had selenium in the jeep and injected it. It had cured a few calves in the last few years. Selenium is sodium selenate and the lack of it in calves is often called 'licking syndrome'. On the third day, everything was normal, the calf was running around and her other calf looked content.

June

The weather has been an absolute treat. It is hard to believe that a few weeks ago the east wind was so strong and cold that in the evening I blocked the keyhole with a small piece of tissue!

Last night as it was getting dark, I looked across to Ard Heiskir and the rocks in the shallows looked bigger than usual and I soon realised they were animals. I concluded that someone must have left the gate open on one of the machair fields with our cattle. This has happened before. Some people know how to open gates but not how to close them. I was worrying about the cattle, and was about to call Angus but had a closer look through the binoculars. I was relieved not to see cattle but twenty-two deer on Kyles Strand. I watched for half an hour and the deer seemed confused and disoriented as they slowly walked towards Claddach Knockline. It was an unusual scene, but the deer better not come near the machair when the corn is growing.

The grass is growing fast and we see early wild flowers – daisies, primroses and birdsfoot trefoil. Corncrakes are back and calling. We have a few calves still to be born, Carianne likes checking them and tags the calves. I think cattle trust women more and develop a strong bond with them, perhaps because men have to do the rotten jobs like dosing which cows don't like at all!

I travel over to Vallay once a week and am happy to be semi-redundant. Carianne shows me pictures of all the cows and newborns cooling off in the sea and eating seaweed. Alexander goes with Carianne after school and at the weekends. He is able to do as much as many men and enjoys the work.

September

Last month saw the first lamb sale of the season held at Lochmaddy Auction Mart. The previous day was miserable with wind and torrential rain so it was a treat for the crofters to get a beautiful sunny day with a slight breeze to keep the midges away.

I thoroughly enjoyed the day even though Paible was the last township to go through the ring. I was keen to wait because this was the first time Fraser and Carianne had sold their lambs at the mart. It was a buoyant sale and I saw many smiling faces. The buyers had more confidence with grazing on the mainland and in the Outer Hebrides improved.

In previous years, one of the first people I would see was the late John Morrison. If it had not been for him during poor sales, many crofters would have gone home with near empty pockets, but he always gave the sales a boost. I was pleased to hear Roddy John following in his father's footsteps whenever the auctioneer lowered the hammer and declared the seller 'Morrison, Lochmaddy.'

I missed a chat with the late Alasdair MacDonald from Ahmor but it was nice to see a large consignment of lambs including some unusual Hampshire ones. I spoke to Mr Shaw who bought from Fraser and Carianne. His grandfather, Bob Shaw, used to buy stirks from my father at the Clachan Sale in the 1960s.

I wholeheartedly agree with the view of Jane Twelves in Am Paipear that sea eagles should be removed from these islands. Without question they are remarkable, majestic birds but their beauty is only skin deep. Last spring a couple were spotted on Vallay near to where a young cow was close to calving. The calf was born alive, but it was found dead soon after its birth with all its organs removed. Crows could not do that and there are no stray dogs on Vallay. It is time for control. Or do we have to wait until a child is carried away?

October

The most exciting day I ever experienced was 7th March 1988 when I sold the first Ardbhan pedigree bull in Oban. I thought back to that day when it was announced that Donald Morrison was to retire from auctioneering in September. Donald knew where the animals came from, how they were reared, and he could pronounce their Gaelic names perfectly. I am sad he will not be in the rostrum. although we are assured he will be working just as hard for crofters in the background.

On that day in 1988, the animal was Pibroch Dubh, a two-year-old black Highland bull. I was disappointed that he didn't win a prize at the show held on the morning of the sale. Before I left for Oban, a manager from Rhum had phoned and offered me £1500 for him, and I refused. When the sale started, the reserve champion only made 900 guineas so I panicked because my bull hadn't been placed in the show. I combed his coat to perfection, took a deep breath and walked into the ring. Donald immediately put me at ease and convinced the buyers that this attractive young bull had the best pedigree having been reared on the wild Hebridean island of North Uist. I couldn't believe it when bidding started at 1000 guineas, rose to 2000 guineas and an Australian took the price to 2400 guineas. Donald held the wooden hammer and I was gob-smacked when it didn't go down. Instead he held the hammer higher and announced in his clear South Uist accent, 'Two thousand five hundred guineas and he is going back to Uist.' Angus and Margaret MacDonald had arrived at the ringside in time to place the last bid. My heart was pounding and tears poured down my cheeks with sheer joy and excitement.

November

I continued to attend the Oban cattle sales and always admired the cattle Tom and Eleanor Muirhead sold from their Corriemuckloch Fold. Each year they sold one outstanding golden-yellow heifer for four to five thousand guineas. Tom always named his heifers Lily Ann. Mr Palmer, an English cattle breeder, bought a Lily Ann and there she met my bull Pibroch Dubh. Several years later I was in Oban wanting to buy a dun heifer but none were listed in the catalogue. I walked round all the pens and there was a dun two-year-old heifer who had been entered in the catalogue as red! Better still, she was the daughter of a Lily Ann and Pibroch Dubh, and was named Lily Ann of Pooks, the fold name of Mr Palmer. Luckily few buyers knew about her background and dun (*odhar*) had become unfashionable, and she cost less than three hundred guineas. I gave her to Fraser for his birthday.

I was shocked to see the disgraceful shooting spectacle in Islay where an American tourist posed with goats and rams she claimed to have shot. I'm not aware of wild goats causing problems on Islay. In my mind, animals should only be killed for food, or for welfare reasons but what was shown on the news was killing for its own sake with the American posing in a film star outfit with her trophy. It was offensive, and not a good advert for a Hebridean island.

December

Rocky, my eight-year-old collie, adores Cissy the cat. Cissy never used to come in the house, but now does, although she's out at night. I feed Rocky about 8 am and 4 pm. Cissy meows outside the door at 6 am and Rocky comes into the bedroom pretending he wants to go out so I will get up – and let her in. I feed her then though I do wonder where she puts it as she steals my neighbour's cat food, and catches rats and small birds. If Cissy comes in when it's time to feed Rocky, he won't go near his food until she goes back outside. Once when I was in a hurry, I put down both dishes and went to get ready. When I returned, there was Cissy with her own dish empty, eating Rocky's food while he stood and watched. Most dogs would growl and chase the cat away but not my Rocky. He treats the hens in the same way.

In the 60s and 70s, our David Brown tractor didn't have a cab but I enjoyed working with it. Recently I sat in one of the massive ones. Cameron MacInnes was driving and towing a cattle float, meeting the Berneray ferry with cattle going to Stornoway. I was impressed with how comfortable it was with a cushioned seat and seat belt! I am used to moving cattle on the road, sometimes on foot, sometimes leading with my wee Suzuki, and that can be a nightmare. When I'm in a car on the road and a heavy vehicle comes towards me, I signal and go the passing place, whether on the left or right, meaning the vehicle can carry straight past me. My recent trip in the tractor sure opened my eyes. On the way to Berneray, only one in about a dozen drivers stopped in the passing place. One came flying over a blind summit, shot straight past the passing place and didn't think to reverse back. I held my breath as we squeezed past, but Cameron is a good driver and he took us up on the verge otherwise the oncoming car would have hit us. It was a risk. The Committee Road is very narrow and the verges are soft and could collapse. Drivers cannot tolerate livestock on the road and can't even tolerate them being driven in a livestock trailer.

Straw and hay are very expensive on the mainland, and on the radio it was said that farmers may have to use sand for bedding. Living on an island with sandy beaches, we have always used sand. I remember twenty-five years ago, there was a special way of placing the sand sods in the beds of the cows. Sandy Humphrey, from Sollas, called and helped me set down the sods. It was funny seeing her struggling with the heavy sods breaking up in her hands and filling her wellies. Sand is good and warm on a sloping concrete floor.

2019

February

In a few weeks, it's the Highland Cattle sale in Oban, and there are new rules. Female cattle over twelve months must have a snitch in their nose – what we used to use on the bulls. Although horns can make Highland cattle look aggressive, they are actually very docile provided they've been kindly handled. Snitches are unnecessary, and look cumbersome and ugly on a female. Bulls used to have a snitch in a show or sale environment, but must now have a permanent ring. We only found out about this rule when we read the entry forms. We had three bulls that had to be ringed and it was a worrying time for me wondering how the poor animals would react but our capable vet, Ealasaid Dick, did a brilliant job. The bulls weren't bothered, though they didn't appreciate the anaesthetic, and were soon back in their field munching on a silage snack.

Why are rings compulsory when they are no doubt uncomfortable in the noses of poor bulls, and they can be a problem. In 2010, we bought two new bulls with rings and were put in our Road Park field. I noticed when I drove past that Erchie had his head close to an old post where a gate had been. I climbed the fence and saw that his ring had caught in the gate hook. I pushed him forward but whenever I grabbed the ring, he pulled back again. Meanwhile the other bull was jumping on his back, annoying him. Young Ruairidh MacDougall drove past and saw my predicament, jumped into the field and between us we got the bull free. If that bull had been out of sight, he could have ripped the ring from his nose. Angus took both rings out the next day.

Imagine driverless tractors and cars! Driving the peat machine is the most boring job so there might be a place for robots there. At least it would not cause an accident. Drones would be handy for checking the machair cows at calving time, and a veterinary robot to stand by to attend to problems. But if all that happened, us stupid humans would have nothing to do.

March

At the annual North Uist Agricultural dinner, auctioneer, Donald Morrison, was the speaker and he was both brilliant and hilarious. I know he was an expert auctioneer, but I did not know he was also a comedian. One lady said her face hurt from laughing. Donald was given a gift, but in the spur of a moment, he auctioned it off and raised £100 for the North Agricultural Society.

Coinneach MacIomhair featured a recording made twelve years ago on Heisker with the late Angus MacDonald Moy who was born there, and passed away recently. It was an interesting, nostalgic talk. He described his experiences so meticulously, and it was clear he had a close bond with the island, and that he loved his life on Heisker.

I wrote last month of the bull Muran Erchie having to have a ring in his nose for the Oban cattle sale. He sold for a remarkable 11,000 guineas, probably the highest priced black Highland bull ever recorded, and a record price for any breed in the Western Isles. With magic social media, news travels quickly, as was the case for the price realised for Muran Erchie. I bet our German farming friends knew it as soon as we did.

At this time of year, the threshing mill would be doing the rounds around crofts in Paibles. When I was young, Donald Allan MacIntosh from Sandary had the mill. Threshing mills were made of wood and iron with a long, wide belt that was connected to a roller on a grey Fergie tractor. This got the mill moving and stripped the barley and oat seed from the straw, much like the mechanism of a combine harvester. Very few threshing mills are around now.

April

In 1944, I was four-years-old and too young to understand what was going on. Almost all the able-bodied men were in the war and only my Uncle Alick and Neil MacIsac remained in the village.

Down on the shore below the house are two large rocks like mountains with a valley between. One day a magnetic mine floated in and got caught between the rocks. A magnetic mine will blow up if it makes contact with another piece of iron. A large heavy square wooden post with iron chains attached had washed up years ago and Uncle Aleck and Neil thought that if they could get that near the mine, and attach chains, they could stop it moving. God works in wondrous ways. Neither of them could remember where the post was although they searched all round the village. Probably it had been used for firewood which was precious with a shortage of peats. How they got in touch with the bomb disposal people I do not know but two men turned up wearing camel-coloured duffle coats. I remember the wooden toggles instead of buttons.

It must have been summer. Neil took the cows to the furthest point of the machair while the rest of us walked to Bayhead to be a safe distance when the mine detonated. My mother and the five of us went to the cottage opposite the school where Kate and Mary Ann lived, and I crouched under the table waiting for the expected big bang.

I remember before leaving the house, grabbing my doll, the first real doll I had ever had, determined to look after her and make sure she was safe. She had a china head and wonderful curly hair. Her name was Hannah and she wore Czechoslovakian dress. I remember all her clothes – a cotton vest and knickers which reached to her knees and trimmed with beautiful lace. Then a petticoat, a white cotton blouse with long puffed sleeves also trimmed with lace. Her soft blue dress was thicker cotton, the bodice sleeveless so the lace of the blouse could be seen. The skirt was decorated with narrow ribbons in red, green and white. She wore hand-knitted red woollen stockings up to her knees and boots of soft, black leather which could be unlaced and taken off.

The entire community was grateful to the two bomb disposal

men for making everything safe. They left the village with gifts of dozens of eggs from the thankful villagers. I wish I had asked more about this when that generation was still alive. Once it was safe, the mine was rolled along the shore and eventually sank into the sand. When I told Alexander a few years ago, he was very keen to move the sand with the tractor and dig up the mine, but I couldn't remember its exact location. It has probably rusted away by now. Before it disappeared, my father removed some metal plates which were heavy enough to hang from ropes to weigh down the corn stacks.

June

Several years ago, Fiona, a Spanish girl, came to stay with me while she was an art student at Taigh Chearsabhagh. She kept in touch and came back to visit in April. The evening before she left, she wanted to cook me a Spanish omelette as a treat. I got a bucket of black peat and got the stove hot. However, just as the omelette was ready, my granddaughter, Sarah, phoned from the machair to tell me my chimney was on fire. I called the fire brigade and a crew arrived within minutes. I had already thrown a bucket of sand into the stove and that extinguished the fire. Nevertheless, within seconds one of the men was on the roof and soon it was all under control. They were so professional and kind, and did their best to keep me calm. Halfway through Operation Chimney, Fiona and I got so hungry we left the men to it and ate the omelette – which was delicious. I felt guilty not offering the boys some but at my age, I have to top up the batteries.

I have eight extra hens and a brand new hen house. My seven old hens sleep in the old house while the incomers have the luxury pad. `So far the old hens do not mix well with the new bunch. Rocky is having a happy, busy time herding all of them. While the old birds retire at 7 pm, the new ones are like teenagers and like to stay up late and Rocky has to help me gather them when I want to go inside for the night. Most evenings two are missing, and when I ask where the last two have gone, he is away like a shot, searching in all the right places. Next minute, he appears with two hens in front of his nose and all I have to do is open the door and in they go, safe for the night.

December

I suffered a stroke on 10th August 2019 and since then life has been very different. A young German girl, Karoline, was staying with me at the time. Had she not been there, I hate to think what would have happened to me.

It was coming up to 10 pm when it happened. Karoline had gone to bed and I was standing at the Rayburn, putting a couple of peats on the stove and closing it up for the night. I remember putting the poker down and thinking, I'm now ready for bed. Next thing I remember is the strength leaving my right leg and being on the floor, unable to get up. I didn't wonder why I was on the floor but I had the sense to shout for Karoline who was quick to check my phone and call the last number. It was my sister, Jessie, and when Karoline passed me the phone, I talked a lot of nonsense to her husband, a mixture of English and Gaelic, and he thought I was a nuisance call. But something made Jessie check the phone and she saw it was my number. She phoned back and spoke to Karoline. Angus and Michelle were in Finland, so Jessie called Fraser and Carianne who sent for an ambulance while they were on their way to my house.

Sarah, my granddaughter, came with me in the coastguard helicopter, helping the paramedics to keep me awake. Why were they all talking when all I wanted to do was sleep? I did not question why I was in a helicopter, nor did it occur to me that I had suffered a stroke. Later I remember being in a room with a doctor and nurse who asked a lot of information. Not once did I think about what was going on.

Sometime in the small hours of the night, I was taken to a bed and had a lovely sleep. When I woke on Sunday, I realised that I was in hospital in Stornoway and that I had had a stroke. Speech was difficult and I forgot words. Movement on my right side was not as it should be. My right leg was bruised but fortunately not broken because I had broken it twelve years ago. The staff were brilliant but I wanted to be nearer home.

I was flown back to Benbecula later that week on an ordinary plane. Carianne had flown up in the morning to fly back with me. I was thankful to be back with the staff of *Odpadal Uibhist agus Bharraigh*[1] with whom I was familiar. I was looked after by kind, loving nurses. I loved listening to one of the cleaners who had a beautiful voice and I always knew it was her passing my door. I became stronger and more normal with every passing day. I had physiotherapy that got my muscles working and my speech came back remarkably well.

I came home at the end of September and have been staying with Angus and Michelle. Alexander is happy to have his granny back home. He visited me at the hospital every day after school and played dominos and made jigsaws to pass the time. It is nice to have company during the long, winter nights.

I have been overwhelmed at the kindness shown to me during my hospital stay and since I came home. I must say thank you so much to everyone, and wish all my readers a Merry Christmas and a Happy New Year.

Note by Iain Stephen Morrison, Am Paipear's Editor

I know that readers will share our delight that the irreplaceable Ena MacDonald has returned for this edition after a time of ill health – it would not have been the same edition number 400 without Ena MacDonald. Indeed it is a mark of her popularity that the question most often put to us over the last four months has been 'when will Ena be back in *Am Paipear*?'

Ena will be embarrassed when she reads this but it seems the right time to remark on what she contributes to Am Paipear. It has not been the same paper without her inimitable take on life, which is real and relatable to the most remarkable cross section of readers, many of whom only read *Am Paipear* for Ena

MacDonald. It is a difficult task running a successful newspaper at this time and her contribution to our thriving enterprise is the reason I have often referred to Ena as "our secret weapon". Her views and stories are enjoyed far beyond the shores of these islands and, without question, she is one of the best newspaper columnists in Scotland.

It was distressing to hear that Ena was taken unwell at the end of the summer, but a blessing to learn she has recovered so well, especially now, just before Christmas. I wish Ena and her beloved family all the very best for the festive season and the New Year.

2020

February

It's almost time for the annual show and the Highland cattle sale. We have two bulls set to send to Oban. Both the red bull and the white bull used to be in the same field but we had to divide them before their play fighting spoiled their *dossan*, the hair on their foreheads. Now there's an electric fence and a road between them, but they can still see each other and are quite happy.

We had Fraser and Carianne's two pigs over Christmas while they were in Gairloch. The red bull stared at the pigs as if he'd never seen any before, and was wondering what they were. At the same time the white bull made friends with two horses to talk to across the fence. Now the pigs are back at the Mill Croft and the red bull seems to be looking for them still.

Did any of you see Apocalypse Cow on Ch 4 showing cattle abroad in huge feed lots where they never see grass and are fed on grain? Everything about their rearing is unnatural. In no way can they be compared to cattle in the north of Scotland that are outside and fed on grass and silage. I wish the media would be more balanced and show more good practices in agriculture and the benefits for wildlife and the environment.

On January 20th I watched the film Beatrix Wood made about ourselves and our cattle. I have been on television before but have never enjoyed the finished film as much as this one. Beatrix filmed it on her own and often in one take. I remember when Nick Nairn came to watch me make black puddings with a crew of four. I had to repeat myself so often. It was hard concentrating while making black puddings and white puddings at the same time because they were going to be eaten in the final shots of filming. I was exhausted. Beatrix came when we were doing jobs and we didn't dress up and it was all very natural. It was emotional seeing the family working together. Beatrix was wonderfully dedicated, with us sometimes at 5 am to get over to Vallay when the tide was out. I received lots of cards and letters.

I remember hearing about an old lady called Rachel who did not go to school but ended up working for the MacDonalds in Balranald. She was a baker and couldn't count so there were marks on the wall for her cooling trays so she could stop when she had made enough. She made the best scones, no doubt with local barley meal and oats.

March

I cannot find the words to describe the weather and can't remember a winter like this one.

Angus let the heifers out onto the dunes on the machair because the field they were in was waterlogged. He gave them silage and they had perfect shelter in the bent grass. However, that was not enough and later in the afternoon they were running beside the tide which was coming in! Angus jumped in the quad and the heifers followed him back to the dunes. What is wrong with cattle that they insist on wandering? You try to be kind, and they go and spoil it.

Last month the annual sale of Highland cattle in Oban was not very good. About thirty bulls returned home unsold. Our red bull, Muran Vallay of Ardbhan got the top price at 6800 guineas and three calves also did well at 1400 guineas for one, and 1000 guineas for two. Two went to the same home in England, and one to Shetland.

Joyce Campbell was the guest speaker at the North Uist Agricultural Society. I couldn't attend the dinner, but I met Joyce with some of our local ladies for lunch at the Claddach Kirkibost Centre. Joyce is involved with the Women in Agriculture task force, and I told her that when I was chairperson for the Crofters Union, I was often the only woman, yet here we were in a room full of women in farming and not a man in sight! Lots of young faces too so the future looks promising.

Lots of seaweed came up onto the shore this winter. Now there are mountains of it on the common so that will be handy for spreading. It's surprising what else gets washed up. A whole coconut came ashore looking edible and healthy. Perhaps it came from the Caribbean.

It will soon be lambing time and crofters must dread the impact of sea eagles on their stock. The Scottish Farmer reports casualties on Skye and Raasay but nothing gets done. When dogs attack sheep there is sympathy for the farmer, and rightly so, but when these vicious birds attack, it's as if it's their right to do so. SNH wants to give money for sheds for indoor lambing but what about the lambs that are taken weeks after they are born? Blackface sheep do not like being housed, and all sheep need to get out to graze. What good do eagles do except attract tourists? Do visitors even know the facts? It would be a good idea to display posters of eagles killing lambs, calves and even dogs.

April
'They're only animals.' How often has that been said?

Last week we had one nice day so I thought that I would go for a walk with Mary Flora and chase the heifers down the machair. Sometimes they decide to come around the shore below the house and then we don't know in which direction they'll go. I took Rocky with me and hung his whistle round my neck. I said to Mary Flora, 'Do you think that I'll be able to walk as far as the heifers?' She thought that I could and if it was too far, I could find a place to sit. I was fine and happy to be going to see the heifers. It's been so wet and windy that it was a novelty to go for a decent walk. When we got to the heifers, Rocky was also happy and went from left to right and back again making sure that he hadn't left any behind. We had turned them around so that they would walk further down the machair. Then two heifers left the bunch and came running over to me. I just couldn't believe it! Here were Tee Pee and Buidheag Carianne, two orphans

I had reared and not seen since last May. Tee Pee started to sook my fingers and it was the sweetest thing that could have happened. I then had a job moving the rest away. I was wearing a different coat to the one I wore when I was feeding them in the summer of 2018 but they still knew me, even though I never called them over.

How things have changed. It's a long time since I enjoyed a good salt herring and a tasty fresh one is even more difficult to get. I remember a van coming round with fresh herring when I was young and we ate them like sweeties. They were fat and Mammy dipped them in oatmeal and fried them in suet. The herring was salted in wooden barrels and every family bought a barrel during the winter.

Recently I heard a fascinating story about salt herring and the best of garden soils. In 1878, Lady Cathcart inherited South Uist Estate. Seemingly she wasn't sure where she would build her lodge so she planted trees at Grogarry and Loch Druidabeag to help her make up her mind. Although her favourite spot was Loch Druidabeag, she went for a walk there one evening and found the midges unbearable so she chose Grogarry. Now at that time herring was so plentiful that Russia imported it from South Uist. The Russian ships came to Loch Skipport and this is where it gets really interesting. Mrs Gordon Cathcart had control of the cargo handling at Loch Skipport. The ships used soil for ballast so it was emptied at Loch Skipport to make room for the herring. From there, the soil was taken by horse and cart to Grogarry Lodge and that's the soil that was used to set up the garden there. No wonder the garden was so fertile. Imagine herring from Uist going to Russia and Russian soil coming to Uist.

May

Very few people make tea with loose leaves these days but I only use teabags in a flask or an emergency. Recently we've been buying milk in glass bottles from The Wee Isle Dairy in Gigha and it's delicious! I haven't tasted milk like this since I milked cows myself so it's a real treat. When we were young, some of the women would keep the last jugful for the tea and the porridge as there was more cream in it. Before I went to school, I walked down to Balemore with my mother to visit Mary Ann and I can still remember the tea that she gave us. It was in a real china cup with beautiful blue flowers and because I was only four, the tea was half milk so it didn't burn me. I had never drunk from a china cup before. I can't remember what I ate because I was so interested in the cup. No-one drank coffee in those days.

In the Scottish Farmer it says that dairy farmers are having a huge struggle because coffee shops, restaurants, hotels and cafes are shut and there is a surplus of milk. Some dairies are even pouring it down their drains, so what about increasing your milk consumption? I drink about a litre a day.

Since I am talking about food, I'll tell you how it was in Inverness when I stayed in Culduthal Girls' Hostel while attending Inverness Royal Academy in 1954. On our first day we were asked if we wanted porridge or cornflakes for our breakfast and whatever you decided, you had it for the full term. I took porridge which wasn't too bad. We walked from school to have lunch and that was the best meal of the day. Quite often we had stew or mince and potatoes with soup one day, and pudding the next. We got back to the hostel at 4 pm and ate at 5 o'clock. One evening we had one slice of white loaf with butter and an orange. I can remember the matron saying, 'You west coasters don't know how to eat an orange.' She went ahead and showed us how to cut the peel into quarters and take it off. Another night we had an apple and one slice of bread. My favourite tea was one sausage with baked beans and the usual slice of bread. I quite liked Dairylea cheese which was also on the menu, but one slice of bread was rather mean, I thought! Then we got supper at about 8 pm and that was the outside slice of the bread. The jam must have gone on at lunchtime and been absorbed because it was dry. Every slice had mould on it but we ate it anyway, and we were given as much milk as we wanted, the one thing that wasn't grudged to us.

I remember the most frightening night in the hostel. Myself and three girls took turns to raid the larder. I was terrified! We took lumps of cheese, bread and apples back to our dorm. I still remember the creaks, thinking it was the matron and nearly collapsing. Every time I hear of a robbery on the TV I remember that night!

I was most frightened to hear about the closure of the schools and definitely we never thought that all churches would close. Religion was not in the equation – every religion has been treated in the same manner. I would like to thank Rev Calum who pointed me to the right chapter and verses in the Bible guide to me – James Chapter 4, verses 13-17. You have to read it a few times to understand it, but the summary is: 'Today or tomorrow we will go into such a city and continue there a year and buy and sell and get gain. Whereas you know not what shall be on the morrow. For what is your life? It is even a vapour that appeareth for a little time, and then vanisheth away. For that ye ought to say. If the Lord will, we shall live, and do this, or that… Therefore to him to knoweth to do good, and doth in not, to him it is sin'.

Congratulations to Lexy Blair who celebrated her 100th Birthday on Monday 27th April 2020. Lexy is the daughter of Ceit Dhodaidh, sister of John MacDonald who lived in Kyles. She is still walking about the house and is looked after by her daughter, Catriona, with her son, Donnie, visiting.

June
I'm sure many islanders are enjoying the Maclean Bakery and Stepping Stone series on BBC Alba. Both stars admit that they make mistakes but only blame themselves. On Mother's Day they forgot to make rolls, so they made scones instead and

nobody cared. I liked the way that they talked about their mothers as extremely capable and never wanting anyone to be hungry. She always told them to make sure that they never ran out of flour! Their flour is the best that you can buy and to my knowledge they have never been short of it.

The calving is doing well and Carianne is very busy. I was thrilled to hear that Lily Ann Fraser had a female Highland calf from our Big Boy. This Lili Ann is the last daughter and indeed calf that Lili Ann of Pooks had before she died. Let's hope that the calf will be healthy and strong.

This is the story of Lili Ann: The Lili Anns of Corriemuckloch were so beautiful and I could only afford to dream of ever owning one. At 5,000 guineas a Lili Ann broke the Highland female record in the early 90s. In 1996, at the Oban Mart, I was looking through the catalogue hoping to find a dun two year old but they were all red. I started wandering around the pens and spotted a lovely dun two year old. On checking the catalogue I could hardly believe my eyes; here in a pen was a dun heifer that had been registered red. Then on reading the pedigree I discovered that her name was Lily Ann of Pooks and she was a daughter of Lily Ann 13th of Corriemuckloch and to my amazement her sire was Pibroch Dubh of Ardbhan. This was my last opportunity to fulfil my dream and I had to buy this heifer. Pibroch Dubh was the first pedigree Highlander sold by Ardbhan and he made 2,500 guineas in 1988. He was purchased by Angus MacDonald in Benbecula for his Druim Dearg Fold and about four years afterwards he was back in Oban making 1,500 guineas and was sold to Mr Palmer of the Pooks Fold in Wiltshire. In 1993 Pibroch Dubh met Lili Ann 13th and the following year she produced a beautiful dun calf. And here I was admiring her at the Oban Mart. The following day at the auction sale I had the first bid and the last. And after a long eight hour journey Lili Ann 13th of Corriemuckloch's daughter arrived in North Uist where she became my grandson Fraser's most cherished birthday present.

Many of you will have seen the VE Day 75th Anniversary celebrations on TV and listened to the Queen's speech. Although the coronavirus restrictions were ongoing, I thought it the best I had ever seen. No-one would believe that the programme was made under restrictions. I was five years old on VE Day and the songs *The White Cliffs of Dover* and *We'll Meet Again* were sung, and seem as popular as ever. It was lovely to see Vera Lynn remembered as well.

It may surprise you that I have had two pet greylag geese for the last three years. I kept them from flying away by trimming the feathers on one of the wings. However, the male's feathers grew long enough for him to fly and one day he took off. The female cried after him but now she is quite happy. The hens go into her pen and keep her company. One day last summer, Sarah put her horses, Mabel and Puzzle, next to the goose pen. I had put their food in the usual place inside their pen and when Mabel saw the grain, she came over hoping to get a mouthful. The geese weren't going to let anything eat their food and when Mabel put her long neck and head over the fence, the geese got angry and didn't budge! They hissed at her and managed to frighten her enough to make her reverse. It was quite funny to see the huge animal reversing in fright away from the geese when they were only the size of her head. Rocky adored the two and when they escaped a few times, he got them back in. Last time I was on the mainland in March 2019 I think they missed me. I went to see them when I got back and called Bic Beag, Bic Beag, Goosey Goosey and they came over and greeted me loudly as if to say 'Welcome Home'. They were never tame enough to cuddle but they didn't run away from me either. Before being allowed out I used to cut grass sods to make sure that they had plenty to eat. So, the next time that I write about the geese, you'll understand that it's not because I hate the geese but because our livelihood suffers.

I wasn't going to plant vegetables this year but Mary Flora insisted that she would do the heavy work. Angus made sure that it was rotivated twice and sea-weeded. We planted seeds for carrots, turnip, swede, cauliflower, cabbage, leeks, onion sets and Little Gem lettuce. The heavy rain and warmth has given them a good start and we'll get rid of any weeds that appear. There's nothing better than your own fresh vegetables, especially carrots. I don't think I've ever been without them.

Hopefully next time that I write, lockdown will have eased and you will have been able to spend time with family and friends.

July

On Air an Lot[1] tonight, poor Donald Sweenie was wondering what was going into his hen house during the day and eating the hens' food. Whatever it was, it wasn't killing his hens nor upsetting them or else they would have been screeching and running away. He happened to be around at lunchtime and could hardly believe his eyes! He saw one of his sheep go into the henhouse through the wee door that the hens use. He followed her in, going through the big door, and there it was having a hen breakfast! She made a dive to get out because she knew she shouldn't be there. They say sheep are stupid but that one certainly wasn't. I don't know how she knew that there was food there but my guess is that she watched the bag go in. Whatever way, she gave me a real laugh!

On Thursday, Alba TV showed the final programme of six episodes about the bakery and Stepping Stones. Many viewers must be amazed at the work and I couldn't get over the number of people they fed during the ordination into the priesthood of Father Ronald Campbell, the first priest born in Benbecula. The first night they served a hot meal for a hundred people, and the next day they served a cold meal for three hundred – the Priest's family and friends. I felt sad when the family talked about retiring from the Stepping Stone restaurant. They have been there a long time and have done their utmost to please their customers and I wish them well.

Since I last wrote, I have been working in the vegetable garden. I noticed that the early carrots are the slowest and the St James Intermediate are the best. They got the same amount of seaweed and were planted on the same day. Everything else is as normal, especially the Little Gem lettuce. They never let me down! Last year, I did everything in the garden on my own but this year I haven't got the same strength. I always used to do the weeding on my knees and it was easy. Mary Flora and I went down on our knees, I was managing fine. It was very warm so when I finished a couple of rows, I sat on the sand admiring our work. Then I went to stand up and I couldn't. We both started to laugh. I couldn't believe I was so much weaker than last year. We stopped laughing and Mary Flora managed to get me on my feet. I didn't want to give up so when I went indoors, I was more determined than ever. I thought if I can get out of the bath without help, I should be able to stand up in the garden without help. At first I had to practise in the house. I went down on the lino floor on my knees and got up no bother. The next day I stayed on my knees longer and up I went and managed to get on my feet before Mary Flora started laughing again.

I have mixed feelings now that the island is beginning to open up again. On the one hand it will be nice to see family and friends but on the other hand it's worrying to think that the virus is still out there. I hope that everyone will be careful and stay safe.

August

If it were not for the virus, we would be preparing for the cattle show this week. I remember taking a Blackface sheep to the show and winning prizes. I also took a fleece. It came from a beautiful gimmer with hair so white and long you could hardly see her legs. I was disappointed not to get a prize. When it was time to remove the entries, the judge was standing near the fleece and without letting on that it was mine, I said I wondered what was wrong with it. 'It's really beautiful, but whoever put it there shouldn't have bleached it,' the judge said. I didn't argue but was very hurt that a judge could say such a thing.

I had a similar incident with butter. I entered salted and unsalted. I had a Jersey cow then and the cream was thick and yellow which made very yellow butter which I decorated with a beautiful three-leaf clover. I didn't get a prize because the judge thought colouring had been used.

Alexander turned fifteen at the end of July. It seems like yesterday I was writing about his childhood. I remember showing him and Archie MacLellan their first chicks. They held one each in their small hands and cuddled them to their cheeks. I still count that article as one of my favourites. I must ask them if they remember that day. They thought they could take a chick home with them but I told them their mother would miss them. That was eleven years ago.

In The Herald, people were complaining about the taste of Ayrshire potatoes compared to years ago when they looked forward to them. Back then, farmers used seaweed as fertiliser but not any more. The secret is to get seaweed on the ground in November before planting in April or May. Two years ago we did that and had a wonderful crop with potatoes without a blemish, nor was there a weed in the plot. Potatoes are my favourite food. Potatoes, salt, butter and a glass of milk!

September
I was going to write this piece in my own words but it is more vivid in Carianne's. She was the one entertaining Emily.

Little and Large
Just a three-year-old girl patting a fifteen-year-old Highland bull, no biggie! Wee Emily has shown an avid interest in croft work since the day she could walk so I told her Mum, Claire, she'd

be more than welcome to be my little helper any day she wanted. So meet my apprentice!

She had an exciting morning that started off in the bouncy telehandler with Angus. Then she joined me on my daily rounds starting with puppy playtime, while I cleaned out their stinky bed, and then she made up the calves' milk. She was particularly amused by Ace, our collie, cleaning up the spilt milk powder and he's now been renamed The Vacuum. She was delighted when the pet calves, Harry and co, slurped up all their milk. 'I made that,' she told them excitedly. We fed the pigs, disposed of some stinky lobster shells, and then we went to collect a takeaway to deliver to Fraser and the building squad. They've been hard at work finishing off their latest house build since 4 am.

On our way home we called in to see Big Boy and his girls. As soon as we were in the field, Emily eagerly asked if she could pet a bull. Big Boy didn't bat an eyelid when the tiny human approached, and as I advised, she chattered away to him as she got close, watching out for his swishy tail as he swatted away the flies. My heart melted seeing her tiny frame next to him! She was so calm and reassured him, letting him know that he's a good boy and that his hair is so soft. Stock lady in the making, I think! Looks like I've found my replacement for when I'm old and doddery. She met my favourite cow, Tee Pee, and gave her a big scratch. After asking what were the dangling black things underneath her, she was amazed when I squeezed some milk out. 'So that's how the milk gets to the shop!' she exclaimed. Nothing gets past this one! Then it was time to go home to tell Mummy all about her big adventure. And next time we're going to meet the even bigger boy, Erchie. Watch this space. I'll be out of a job at this rate.

On the 13th July, Fraser got an email from a lady in Australia who had watched the Rick Stein show. She searched and found the film made by Beatrix Wood for BBC Alba, and then wondered if the Ena MacDonald was the same one she had met on her family farm in 1962. I could hardly believe my eyes when I read her email. Their names didn't really mean anything but the name of their farm did. She was then twelve-years-old and I was twenty-two. I asked Carianne to email her and get her address and telephone number. It was wonderful to talk to her. We spoke about the farm with a small number of Jersey cows which I helped her mother to milk. I told her that I could remember the place where they milked and how the door was set up. Seemingly I made stovies[2] there and she never forgot the taste. She said that I melted butter in a pot and then added whatever vegetables they had. I still make them and they are cooked very slowly in their own juices without any added water. She told me that her parents retired in 1980. Her Dad died in 2000 aged 82 and her mother bought a small place near her other daughter and passed away in 2017, just after her 100th birthday! I looked in my old address book and their name was at the top for 1967 and 1968.

It was a lovely place for a farm. My husband and I had taken our bikes and gone camping north of Sydney. Heavy rain came and we went to the farm and they made us stay for a few days. We visited them again and I kept in touch after I came back to Uist. I don't know why we lost touch but now she has made contact again, we will continue to correspond. What a small world it is!

My sister, Flora, in Renfrew, has a hearing aid and finds it very helpful. A few weeks ago she was walking home from the park with her little dog, Shona, when it started to rain heavily. She remembers pulling up her hood but doesn't remember where. When she got home she discovered she had lost one of her hearing aids, probably fallen out when she pulled up the hood. A few days afterwards, she was walking down the lane and she saw one of her friends with his dog. He started talking to her but she could hardly hear him. 'Oh.' he said. 'I have one in the house that my dog found in the lane just about here. I took it from him and put it in a safe place. Wait a minute and I'll get it. Would you know if it was yours?'.

And she laughed saying that would be too good to be true! Anyway, back he came and with the hearing aid that she had lost!

His dog had picked it up and had gently given it to his master. How strange that the man had spoken to her that day and she had mentioned her hearing aid and he had it safe in his house.

Last week I was waiting for Michelle to come and feed the hens and she was later than usual. Rocky loves going with whoever feeds them and he was getting restless. I thought that I would test his intelligence and I said very clearly to him, 'Go and get Michelle and tell her to give the Bic Bics their food.' I said it twice and he ran to the door. I let him out and off he went. Michelle arrived a few minutes later saying, 'I think that Rocky was trying to tell me something.' It was a beautiful day and she had been sitting on a large stone having breakfast, looking at the sea. He put his nose under her arm and tried to make her get up. He followed her in and was pleased when she grabbed the hen's food basin. I sometimes think that he understands every word that I say. Such a good dog. I would be lost without him!

The islands are busy with tourists and family friends, but many of us are still quite nervous that some will bring the virus. I'm glad that the schools are open again so I hope that the children will be careful.

October

I used to have Jersey cows and they were a treat to look after. Kalmia was the first. She came as a calf from Dr Julia in 1968. Angus was only three years old and Kalmia was bought for £18 with cash from his piggy bank. I wrote about Kalmia years ago, how she arrived in John MacBain's post office mail bag. Can you imagine that happening now? Kalmia had a heifer calf which I called Julia and she calved in the winter in the byre while we had deep snow and frost. I thought the calf was cold so I took it into the kitchen beside the Raeburn. Rocky the 2nd slept in the back porch on top of a warm woollen red coat. When Rocky came in, he saw that his bed was underneath the calf. I can still remember how Rocky stared at the calf as if wondering how he could sort out this problem. Next minute he grabbed the coat in his mouth and pulled it. Fortunately it stuck around the calf's hoof. Rocky stood back and wondered what to do now. It didn't take him long to make a decision. He lay down on what had been pulled out from under the calf. He was tired and all he wanted was peace and a warm bed, so he put his head on top of the calf's back as though he was cuddling her and slept. I thought it lovely, how Rocky treated the wee calf. Anyway, when the calf woke up she was warm so was returned to the Byre and Rocky got his bed back.

Last month I wrote about making stovies in Australia and since then I have heard that others have been making them. My sister, Jessie, makes them regularly. She always gives me a row for not eating much fruit but I tell her that I have a very healthy diet of our own vegetables fertilised with seaweed. We have potatoes (my favourite), carrots, onions, turnips, parsley, chives and lettuce. Flora hadn't made stovies for years because the potatoes that she had bought were really bad so she didn't want to waste the other vegetables. Without decent potatoes you can't have good stovies. However, last week she got a big surprise – a neighbour must have read her mind – a knock at the back door and there he was with a bag of potatoes grown somewhere near a golf course. They were huge and dry and called Finland's Finest Chippies so Flora has good stovies now too.

My letterbox, made by David, my brother-in-law, many years ago, was leaking during heavy rain. As soon as he heard, he was on the job and took it down for it was rotten and made a new one, painted it red, and it's back as a landmark with Ardbhan on it like before.

It's frightening to hear that we now have the corona virus in Uist. We were all shocked when the news came. It's at the south end but this island is small and it will be difficult to avoid each other. There are rumours about how the virus came to Uist but one thing for sure is that some people have broken Government guidelines and the rest of us will suffer. We were all so thankful that we had not had any of the virus here until last week but were we thankful enough? Psalm 107 is well worth reading. It would take up to 10 minutes if you have time. Verse 19 says, "in grief they cry to God, He saves them from their miseries," and then verse 31 reads how we should praise and thank the Lord.

2021

February

The Monarch Islands belong to North Uist Estate but Roderick Kirkibost had a lease and used to graze his cattle, probably bullocks. Whenever he wanted the cattle checked, he took a few strong men with him. In 1953 those men were my own uncle Hugh, Uisdean Eoghainn, Andrew Laing, Calum Dhodaidh, Archie Morrison and Roderick Kirkibost himself. Their boat was an old lifeboat that had been used in Barra. Their trip to the Monarch Isles gave them a good opportunity to shoot some barnacle and greylag geese as well as cormorants. We did not have a surplus of geese in those days!

A few strong waves hit their boat just past Beul a' Chaolais making it leak slightly but they thought it safe enough to carry on to Heisker, get her out of the water and mend her before their return journey. They left here on a Wednesday and couldn't return after mending the boat because there was a mist. When they still didn't return when expected, the alarm was raised and the new Barra lifeboat went out to find them. It travelled round the island and reported that there wasn't a soul on Heisker. A plane joined the search and it too reported seeing no-one, although we found out later that the men had seen the plane. Everyone at home thought that they had surely drowned and there was doom and gloom in this village and in others. I remember my father had been down at Arnol Point looking through the telescope but there was no sign of them, and he was sitting in the kitchen with tears in his eyes. Seonaidh Dhodaidh in Kyles had the only phone in the district, and he talked to the press about the situation.

I was thirteen years old and I kept going out to the back of the house to look for them. My diligence was rewarded and I spotted them between Kirkibost Island and Kyles. I ran in to tell everyone and they were so happy! Then I ran up the hill and told my two aunts that they were safe. Of course I then let everyone in the village know. We found out later that Alasdair Laing had also seen the boat when he was out hoping for his father's return.

Alasdair told me later of a song that his father had composed about the experience. Archie Morrison was reported as using sea water to cook potatoes as they had forgotten salt. The men had enjoyed being out in Heisker and did not know that there had been a search party out looking for them.

After a few days, Seonaidh Dhodaidh got a newspaper from one of his brothers living in Glasgow and on the front page was the headline: ENA, THIRTEEN, SEES BOAT. That was the first time my name was in the paper! It was so exciting because the islands were hardly ever mentioned.

After this, Roderick took more men and usually two boats, one pulled by the other, possibly a barge. The cattle had to be tied by the neck and legs which is why he needed strong men because the cattle were huge from eating grass and seaweed. My Uncle Hugh was probably the best shot amongst them. He had orders that if any of the cattle broke free, he wasn't to wait but to shoot it immediately. His rifle was a .303 calibre and he had to act quickly because if a cow was moving about, it could go through the side, or worse, capsize the boat. Thankfully I don't think that this course of action was ever needed.

March

Fraser and Carianne are building their new house. They had to do landscaping and then laid the foundations. They are having underfloor heating installed. I'm afraid I don't know how that works! The trusses arrived on the evening ferry and the next day everything was up. The kit was made by Fraser and his team in the shed. They have the sarking[1] already in place and much of the block work. I keep looking across the water with my binoculars and am very excited to see it all going up.

The Dusary Mill tenancy has been in the same family for years. The miller was married to my great-aunt Maggie. My sisters and I used to walk across the ford when the tide was out and played with Peggy at the old mill. The water, turned in the big wooden

237

wheel, came from three lochs. The furthest was at the Committee Road on the right before you went over the cattle grid. You can see the remains of the dam that stored the water in a dry summer. Barley was the main meal but they might have been making oatmeal too.

The Mill Croft doesn't belong to a township. The small field across the main road where White Rock Cottage is, was used by the mill tenant but belonged to Kyles Paible. It was probably a verbal agreement to give the horses that came with the seed some grazing. One of the walls supporting the mill was still standing until 11th January 2005 when it came down during that night's terrible storm.

Lasair Chlach Chanaidh, the Flint Stone of Canna, was a black Highland cow bought from Canna. She was lovely natured and one of my favourites. In the 1980s when the rest went up to the ranch, she was left behind with the Jersey cows because she hadn't calved. However, poor Lasair didn't like that and escaped to the hill to join the rest.

I was up there ten days later and she had just calved, but the newborn had fallen into a water hole and drowned. She stood there, not wanting to leave her calf. I took it out of the water, and took Lasair home as we had four orphan calves and I'd get her to suckle one. I phoned Iain MacQuarrie and we met on the Committee Road. I had dragged the calf down the hill called Marrival and Lasair followed. We got the two of them in the box. She was full of milk and let me milk her. I had a live calf in a pallet pen by the gate and to my amazement she let the calf suck but wouldn't lick her. She was a strange cow. For three weeks she allowed the calf to suck but only when I stood there. She never showed the calf any love.

Then we made a decision. We couldn't leave her like that all summer and she would need to be with a bull so we put the calf back with the orphans and the cow walked across the strand with cows going to Gearraidh, She went happily at first but when she reached the grass on the other side of the bay, she ran like a deer back to Kyles. No-one could catch her. She went back to the field she had shared with the calf and cried loudly. I got the calf and opened the gate. She made a dive for it and licked it as if it were a newborn. It was strange how she acted that day. The calf was happy. At last it has a real *mamaidh*[2] who loved him dearly. A week later, she took the calf and joined the others. There wasn't a happier pair on any croft or farm in Scotland.

April

The cattle were inside all winter when I was young and there was plenty of dung around. Work like filling carts was done by hand using a four pronged grape. The ploughing was done by horses and many a back breaking day, I followed the plough, turning the furrow back! The seed was planted by hand, approximately eight large buckets per acre. It was then harrowed, again with the horses. The harrows were made of wood with iron spikes to cover the seed. The birds had a great feed stealing the seed so the sooner it was covered the better!

The sheep were banned from the machair on the 15th May because there were no fences. The crops were then safe to grow. Mind you, some areas had too many rabbits, although many were trapped in the winter and sold to the mainland, as well as being our staple diet. There weren't geese to plague us. Only a small number came in October and went north in May. The lambs were born about the end of April, later than now. Calves were born around the beginning of May and were inside with their mothers until the end of May which meant that we could get the milk easily.

Potatoes were planted in April/May and seed potatoes had to be cut in half before planting. Women without their own crofts, called cottars, went around the crofts doing this job which guaranteed them enough potatoes in the winter. I used to think that they were ancient, many widowed and dressed in black. Us children would be in great demand for planting potatoes because people planted large crops to feed man and beast.

Three little girls barefooted and clutching a Bible, each on the way to school. That was a common sight in the late 40s and early 50s. I remember because those girls were myself, my sister, Agnes, and my neighbour, Peggy. Between Easter and the end of June we went to school barefooted. As we walked, we recited the Bible verses and psalms we had to learn for the Bible period in school. We recited The Lord's Prayer every morning. Being young and naughty, we had different versions, by whom I don't know! One was:

'The earth belongs unto the Lord and all that it contains
Except the Outer Hebrides, they belong to MacBraynes.'

As we got older, we never forgot the Bible and it was nice to sit in church and listen to the Minister using the verses we had learned. As I get older, the verses that I learnt in school will always be my favourites. I was thinking a while ago that it would be nice if some of my readers would memorise some of my favourites and teach their children or grandchildren one or two every month. I think the most well-known Psalm is number 23 and one of the best known Chapters is First Corinthians, Chapter 13.

May

The peat machine has been busy since the 13th of April. It's a great piece of machinery but very expensive if it breaks by hitting a stone. It cuts into the ground and takes the turf with the peat and that's why the peat is not as hot as that cut manually with the *treisgeir* – peat spade. Before the machines, every croft had a *treisgeir* and cutting the peat was highly organised. Here in Kyles, it was usually done by men. It's difficult to explain. One person uses the peat spade to cut the next layer down below where the turf has been, walking backwards along the bog. Another person, who is lower than the cutter, catches the cut wet peat and throws it up onto the side of the peat bank. There were six working crofts in Kyles so three pairs of men who cut enough in a day to give one croft a year's supply. It was then up to each crofter to dry their own. If the weather was good, it would take about ten days to semi-dry and the bigger the family, the quicker the job was done! They then put eight peats together in a *rudhain* – one peat standing on its thin, long side, four peats around it, two each side, then two at the ends and one balanced on top. These were put in neat lines. The next stage, when the peats in the *rudhain* had dried further, was to make a big heap on dry ground to allow the air to flow through. The next stage was to get all the peats to the roadside. To be continued next month...

Rachel was a lovely black Highland cow bought from Canna. The late Dan MacKinnon, Canna's stockman, wanted rid of her because she nearly always lost her calves. Her udder was huge and low. We had about twelve cows and we wanted to increase our herd because the prices were fantastic if you could export. Rachel was not tame but Angus and I managed to get the calves born while with us to suck, so they survived. One year she was empty and because she needed milking, we left her calf with her instead of weaning it. Her calf was just one year old and usually a Highland calf doesn't come into season until it's much older. However the bull, Neilie of Auchnacraig, must have jumped the fence at night. We didn't see him serving her. Winter came and went and in April, Rachel was pregnant and unfortunately so was her heifer. We kept an eye on both. I went over every day, and then I realised that she had calved, but there was no sign of it, and she was just licking her heifer. Anyway I found it wondering about, covered in mud she had thrown at it to keep it away from her udder because she still bonded with her heifer. Her body language told me she didn't want it and she was showing aggression towards it. I pulled the calf back a bit and she went back to licking her heifer as though she had just calved her. Angus and I got the calf to the pens, made the calf comfortable with pallets, and then went back and took the other two to the pens. We slipped out the heifer and put her in the old barn. Rachel was now in the same pen as the calf but it was protected by the pallet and we hoped that she would start bonding now that the heifer was away. We left them, and when we returned it was a beautiful sight. She was licking her calf and not thinking of her heifer, so we pulled the pallet away and let the calf out and there they were, mother and daughter, the way they should be. The calf was very big and very strong and very hungry so she managed to suck her without help. It just shows that it isn't wise to leave stirks with their mother or they may get priority when the mother calves. In case you were wondering, wee Rachel calved a lovely male calf on her own which she reared so well that it was the best bullock we had that year.

Last week I was in the local shop looking for a birthday card for a very special lady but there were no cards above 100. I said that I would buy one of those and add one. It is for Lexie Blair who lives in Newton Mearns, the late Ceit Dhodaidh's daughter and sister of the late Seonaidh Dhodaidh. She must have happy memories of her childhood holidays in Kyles because she still talks about the machair potatoes and the black and white puddings. I wish her a very Happy Birthday!

I wrote that last month I would let you know which Psalms and Books of the Bible that our class learnt at school. This month it is Psalm 121 and Ecclesiastes Chapter 3, verses 1 to 10. I hope you will enjoy reading them.

June

Last month, I described the first stages of peat-cutting as I remember it from my childhood, before we had a tractor to take the finished product home! I will finish the description here. We got our first tractor in 1954. Before then we used horses. When the peats were ready, they were carried to the roadside at Committee Road. Most evenings we cut sand sods on the machair and take them to the peat track the next day, if it was dry. My father and brother put the peats into hessian sacks and once they were within throwing distance, they were thrown straight onto the ground. We also had a hand barrow to help carry them.

We built a fire with dry caorain (black peats) and collected water from the steam to boil a kettle for tea. This was fresh and untainted, unlike the hard, limey well water. We packed a picnic – hard boiled eggs, scones with homemade butter, crowdie, cheese and cooked sea trout. We all loved the picnic on the hill.

It was a short day, timed with the tides. We would head up before the tide was fully out and make our way back after it had turned but the water was still shallow enough for us to cross safely. The last load would be topped up and taken home with us walking beside the horse and cart. Once the peats were home, the final job was constructing the peat stacks. That was an art, and there were beautiful stacks outside all the homes back then.

Watching Fraser and Carianne's house being built makes me wonder how the older houses were built. They were made of stone, but where did the stone come from? In Paible in the 1920s, thirty houses were built with slate roofs because traditional thatch was less popular. The walls are 2.5 feet wide so I don't worry when the Atlantic storms hit. When I had improvements done in 2005, the wooden floors were as strong as the day they were laid in 1933. In 1945, most of the byres and steadings had thatched roofs but I only remember one thatched house in Kyles. I remember putting a fresh cover on ours with my father. I spent hours and hours laying all the bent grass in one direction. We found bees on the barn roof which meant lots of honey. In Bayhead, there were seven thatched houses but they were abandoned when the cottars who lived there died.

In 1967, the circular road from Kyles to Paiblesgarry was upgraded with tarmac and stone was needed for that. They took away the empty thatched houses in the area, and the old steadings and yard walls. Many workmen came to our garage for lunch. My father put down two planks for seats and would take out a huge teapot. They had their own sandwiches. My sister, Agnes, was here with her daughter, Morag who was two, as was my son, Angus. The two toddlers would go out with my father and the tea and stand in front of the hungry men to get biscuits. You would have thought they were starving! The men got a good laugh out of it.

It's been a cold spring, probably the coldest May I remember since 1986. That was the year that Chernobyl frightened us all. However, we must not complain as there is so much to be thankful for. The reading for this month is Isaiah 53 and the Psalm is 84.

July

Rocky adores my six hens. He loves to be there when they get out in the morning and when they go in at night. Anyone with hens will know that with the long daylight hours they don't go in till about 7pm. Sometimes two decide it's not time even then and disappear down to the shore. I tell Rocky to go and get them, and he knows where they're hiding and gently brings them up as though they were lambs. Recently I was at that job when Fraser came over to see me, so I said to Rocky, 'You'll be able to get the naughty two in the shed pen along with the two bulls.' I was just joking of course. Out we went and there was Rocky, on his own, pushing the hens along with his nose. We pushed them towards their house and then got them inside. He's a gem of a dog.

Some of you will remember seeing on Alba the film Crodh Gaidhealach an Aird Bhain, The Highland Cattle of Ard Bhain, filmed professionally by Beatrice Wood, a film maker who lives in South Uist. It has been nominated in the feature documentary category at the Celtic Media Festival and has a good chance of an award. I can honestly say that it was the best I've ever seen on Alba. You can accuse me of being biased but given the number of messages, cards and letters I received, others thought the same.

Many of you will know that I had two pet greylag geese. One managed to get out of the pen and never came back, but the other one, Bic Beag Goosie Goosie, was still here. Until last week. I was very fond of him but felt guilty about him being lonely. The hens went into his pen and were his only companions. When wild geese flew over he would shout to them and that made me feel worse. I phoned the local gamekeeper for advice and he said that he would be happy to take him. I knew he had one goose. About a year ago, when I saw him in the shop, I had asked after Sir William.

'Well, I don't have him anymore.'

I was about to ask what had happened when he smiled.

'Sir William is sitting on eggs and is not Sir William anymore. He's now called Harriet.' I had a good laugh about that. Anyway, I took Bic Beag to his new home with Harriet as his friend. Seemingly they had a long goose conversation when they met and Bic Beag was following Harriet all over the place! I'm glad I decided to give him to Colin and I can visit the goose and see if he remembers me.

Our cows and calves were moved on Wednesday 23rd June from Vallay to their summer grazing in Griminish and to the Committee Road. They will be there until October when they will make the journey back to their winter quarters. I miss those trips, but I am not fit to do them anymore. They knew my voice and it was lovely to drive ahead of them as they followed. However I'm very thankful that my family is interested and I know they are looked after and loved.

There are only four chapters in the Book of Ruth. It is a lovely story but sometimes difficult to understand. Chapter 1, verses 16-18, have always been well read. I have chosen Psalm 118 this month, verses 29-32. It is special in Gaelic.

August

Many of the elderly will remember Tommy Darkie and his band. His real name was Johnny MacLeod and he played at John Angus MacDonald and Flora's wedding, and very few missed a dance. Johnny sadly died in Australia this June and will be sorely missed. In the late 60s he came home to Lewis and built a house, intending to stay, but the warm climate of Australia lured him back. When I went to Australia, I met Johnny for the first time at a Scottish ball. In the interval, someone told him that a girl from North Uist was here and afterwards, when he played Gaelic songs from Uist, my homesickness got the better of me and tears blinded me. I was invited to his home where I met his kind wife, Rita. Johnny was great for jokes. He must have had about three thousand written out. One went like this: An old lady approached him and said, 'I love your music, Johnny, and I've been told you play by ear.' 'Oh yes,' says Johnny, 'but I also use my fingers.'

The Green Party blames farmers for contributing to climate change. They advise us to cut about 200,000 cattle in Scotland to cut down on their methane emissions. To put it bluntly, they say that cattle farting is responsible so we must cut our meat consumption. But the supermarkets would just buy beef from other countries who don't have our excellent welfare standards. Already, because of Brexit, we can't export live pedigree breeding cattle.

The sea fords are changing but they always did change. People used to walk to Heisker. And at Baleshare in North Uist, the west side is now under sea. Causeways built since the 60s with insufficient openings for the sea to get through hasn't helped the situation.

I must apologise for my little mistake in the Bible reading last month. It was Psalm 118 and verse should have been 19 until the end (it used to be my late brother Ewen's favourite). The second Psalm is 107, verses 29-32. I will leave it at those two

Psalms to concentrate on as when you read those verses you will want to read them all.

October

Cathy Laing, a local resident in Claddach Knockline, is suffering from pancreatic cancer. She was on Kenneth MacIvor's radio programme a few weeks ago and told us about her illness. It was very moving when she said the doctors gave her so many weeks to live, but already she had passed that time. She said, 'Our Creator didn't tell me that, and only He knows how much time I've got.' I think listeners shed tears that day. Two weeks ago, with the help of her family, Cathy organised a three mile walk on the Committee Road to collect money for cancer research. She could only walk fifty yards yet here she was tackling three miles! Those of you who know the Committee Road will know the passing and parking place near Marrival where the walk started. There must have been forty walkers there. The walk was along the road for one and a half miles and then back. Cathy walked every step and watching her leading the crowd with her white flag, you would never dream that she was so ill. And who joined us after driving from Stornoway and catching the ferry but Kenny MacIver and Iain MacAulay, a lovely gesture of support. Fortunately the weather was good and dry with a slight breeze. Kenny said that when Cathy was on his programme, the girls on radio duty were in tears. The walk was very successful and with the help of the local shops she collected over £6,000.

On the Gaelic news on 22nd September, SNH (Scottish Natural Heritage) now NatureScot were asked if they were going to fund another pilot scheme to reduce the number of greylag geese. They said not but would give free advice to crofters and land managers. Do they really think that advice alone will reduce numbers? For over twenty years I have been involved with the Goose Committee, Crofting Foundation, and privately to help find solutions. I am disappointed that we seem to be no further forward. It's the same carry on with sea eagles. They expect crofters and farmers to get rid of their sheep or build sheds to winter their sheep and lamb them indoors. Don't they realise that those cruel claws can lift a half-grown lamb? Come on NatureScot!

My nephew, Calum, died suddenly from ischaemic heart disease on the 11th August. He left behind his father, his wife, Morag, their two sons, five grandchildren and his two sisters. He was only 57 years-old. Calum was a real family man, a chef and always busy helping the family. He was not overweight and never complained except of a sore back occasionally. Family and friends were shocked as we loved him dearly. When we got the big storm in 2005, the sea came into my house and there was a terrible mess of seaweed, marram grass and debris outside. Calum had a week off work so he and his Dad worked solidly with two wheelbarrows and cleared the lot. That was so like him. He and his family came nearly every summer and in the 70s he went to school in Paible for a few months. I miss his regular phone calls and find it hard to believe that Calum will not be returning to his beloved Uist.

I have chosen the following scriptures for this month – Psalm 121 and Matthew, chapter, 27.

November

Now that winter has come, the cattle are nearly all on their machair pasture where they get a dry bed and shelter from the fierce winds. Vallay Island is fortunate to have shelter from all directions. On the 18th September, all our bulls were taken off the cows and brought to Kyles machair to a small stock proof park which isn't grazed in the summer. The bulls seemed to be friendly enough with each other. Once the cows and calves are off the hill parks, the bulls are then moved to their heather pasture for the rest of the winter. Of course, the potatoes had to be lifted in Vallay before we could put the cows and calves on that grazing. It's had no fertiliser except seaweed and dung from the cows grazing there the previous winter.

Four years ago we had a very premature calf who couldn't even walk. He was tiny and had pneumonia and then had joint ill which is often the result of an infection of the navel because a calf hasn't had enough colostrum. Carianne and Fraser were away and Margaret Cameron was looking after all the pets, sheep and pigs. When I went over to the shed at the Mill Croft to check on the animals, I saw that the premature calf had deteriorated. Margaret had enough to look after so I took the sick calf to Kyles. I gave him warm milk during the day and during the night. He began to recover, and the day he ran across the shed floor I was delighted! After a few weeks we gave him to the Riley family in Paiblesgarry after they lost one of their cows and her poor orphan calf needed a pal. Jacqueline decided that he would always be a pet and never end up in a freezer because he was such a character. She called him Piglet and he knew his name. She moved him and his pal to Aberdeenshire for a while where they were kept at her stable yard. There, Jacqueline managed to get a halter on him and the children who came to the stables had him as a pet.

A few weeks ago, Jacqueline arrived back in Uist and rang me to say she was on her way and had something to show me. I was about to have my dinner and wondered if it couldn't wait until the next day, but she was insistent so I agreed. I made a few guesses about the surprise but they were all wrong. Off to Paiblesgarry we went and driving up to the house, I noticed a lovely bull in the near paddock.

'When did you bring him?' I asked. 'And where did you get him?'

'Don't you remember him?' she said.

'He looks like the Ardbhan breeding but I can't be sure.'

'It's Piglet!' she said.

I couldn't believe it and was excited to see him back, though I did feel a little silly that he wasn't a bull but a bullock. Obviously I hadn't been looking in the right place! His horns were a lovely shape, better than on some bulls at the Oban sales, and he was in lovely condition and very tame. And so my patience at the very beginning of his life had paid off. I could hardly believe that the tiny calf that I had brought back to health could grow to this size!

Rocky and I feel like we've been in the wars these last few weeks. I've had a fall but luckily didn't break anything and poor Rocky hasn't been well and we're still trying to find out what's bothering him.

December

Rocky was eleven years old last summer and although I dreaded the long winter nights, I had him to keep me company and could talk to him anytime. A few months ago, I took him to the vet because his mouth was bothering him. He was slow eating his dinner and I gave him Wilson's dog food with a raw egg and lots of warm milk because he didn't like anything hard in his mouth. The vet discovered two broken teeth and gave him antibiotics. A number of weeks later, I noticed lumps in his throat, one the size of an egg so I took him back to the vet, and she sent away a biopsy. The result came back with a diagnosis of cancer. I was devastated because I knew that there was no cure. He lost a stone in two weeks. His diagnosis together with extreme loss of weight and increasing tiredness meant that I would have to make the difficult but kind decision to have him put to sleep. He began to wake up during the night, needing to go out, and he seemed unsettled, so I knew that the time had come. I felt that he thought it too. On his last day, he seemed very happy. Euna Westford was with me and she took him for a few short walks down to the beach where he played with seaweed. My heart was broken knowing that he could still be happy outside, but in another two weeks he would be so thin, and one thing was sure, I did not want Rocky to suffer. When Michelle came over, she took Rocky to the post box with her and around the village whilst waiting for the vet to arrive. Those moments were hard! Some people might say that he's just a dog but Rocky wasn't just my dog, he was my friend and pal, a loving pet who loved everyone who came to the house. As my readers know, he adored the hens and spent hours with them every day. He loved to get them in at night and out again in the morning. When I had chickens, he was gentle with them and made sure nothing hurt them. If someone knocked on the door he would bark, but if they just walked in, he would welcome them with his big paws in a big hug. When I drove to the shop, Rocky came with me and the boys working there talked to him through the car windows. Rocky was so special. He loved the TV, especially *Air an Lot* and the Gaelic sheep dog trials.

When the time came, Michelle sat with Rocky's head on her knee, holding him gently so that he did not have to go through being sedated. The two Emilys from the vets were gentle and kind with him and he passed away peacefully with us all around him, knowing that he was truly loved. Angus buried Rocky near the house beside another pet dog called Cola.

The house does not feel the same but I still have Cissy, the mother cat, who is a great hunter and was always fierce when looking after her kittens. She has terrified many a dog who got too close! She and Rocky were great pals and often slept beside each other. Now she's using the stove for warmth and enjoying more time on my lap. Rocky used to be jealous of that.

I have chosen verses that some friends suggested for this month, Psalm 19, verse 7 and Psalm 119, verse 116. If you want to read more, look at Luke, chapter 1, as this is very relevant for this time of the year.

I wish all my readers a very blessed Christmas, a happy New Year and a healthy 2022.

2022

February

Saturday 15th January was a very busy day here on the croft and more so on Vallay. It was a calm and dry day and Angus took me for a run in the jeep to collect the last load of calves. Every calf over there was brought to Ardbhan so of course all the cows had to be brought into the walled garden where, with the help of pens, they were separated. They were nearly all eight months old and had been naturally weaned already. I hadn't been over for months so I enjoyed the outing. It was good to see Angus working alongside six young people – Fraser and Carrianne, Craig, Ryan and Orin, and Alexander – all good stock people, all enjoying their work and keeping the cattle calm. It's good to know that there are keen young crofters here still. Now that the calves are home, the young people are busy feeding, cleaning and caring for them in the shed. It's nice that I get to see the calves in the pen every day from my wee window.

The North Uist Agricultural Society couldn't have their tractor rally last year, nor in 2020. It seems that Margaret MacLean (Mags) had a brain wave and asked crofters on social media if they wanted to hold one on Christmas Eve and give people a real treat. The response was good and instructions were given for all tractors to be ready and waiting to leave Hosta at 6.30 pm. There were great preparations with the tractors lit up with Christmas lights. I saw them driving along Ardheisker and as far as Carinish where they turned around, headed to Clachan and then to Lochmaddy and back to Hosta on the north road. From my window with my lights out, I could see them turning at Carinish. You would think that the tractors were dancing like the JCBs did many years ago at the Highland Show. Amazon did a great trade in lights, in fact they were sold out of some because the tractor drivers wanted their vehicles to be as bright as possible. There were about twenty-four tractors and Clachan had the biggest crowd. Santa Claus found time to come in his sleigh even though he would have been busy that night, so I hope that the children enjoyed seeing him. There were also actors dressed as Santa Claus driving tractors and Rory Glebe had a hay bale beautifully lit up on the back of his tractor. North Uist had never seen such a bright Christmas Eve before.

More than two thousand years ago, a star shone from the east and gave the shepherds enough light to guide them to where Baby Jesus lay in the manger, in the stable, because there was no room at the inn. The tractor rally also brought a lot of happiness all over Uist. It was a real novelty and I was thinking, wouldn't it be wonderful that as a result there will be a great Christian awakening in Uist.

April

I often wonder if anyone ever wrote a book or diary about Lochmaddy Hospital. In May 1940, I was born there and in 1966, my sister Agnes's daughter was born there and named after me, Christina MacDonald Ramage. Dr A. J. MacLeod delivered both babies. We used to call that building *Tigh na Bochd* or Poor House. It had approximately forty patients and twenty staff. The work was heavy, very different to the modern hospital. Mary MacQorquodale would milk seven cows every morning as well as working in the laundry. They didn't have washing machines then so every item was washed by hand, cleaned with a washboard, put through a mangle, then hung outside to dry. It was a happy community hospital and it was sad to see it close its doors in March 2001. Morag Sheumas was the last patient to be cared for there

I spoke to Mary Ann Campbell who cooked with Morag Sheumas, Susan Dingwall and Nan (Murchadh Theorlach). They started at 7.30 am and breakfast was usually porridge. Fit patients walked down to the dining room. Dinner was at mid-day, high tea at 5 pm and every Sunday there was a lovely roast. In the 1970s, milk was bought from Alastair Ath Mòr and eggs from Johnny Ban at Clachan Farm. They got their meat from the late Iain Morrison, the butcher, whose slaughterhouse was in full swing then, and he sold meat all over North and South Uist. We grew about one acre of carrots – James Intermediate – in the early 50s. I remember weeding and thinning them with my sisters. Angus MacLellan from Bayhead would take bags of them away in his lorry but I can't remember if they were sold to the hospital. I was told that the patients tended a garden of vegetables too. It's good to see the MacInnes Brothers renovating and developing around the old hospital.

The weather is much kinder now. The constant rain has stopped so the animals must be happier. Many a night this past winter I lay awake listening to the rain pounding on the roof and was thinking of the animals lying underneath it. The ground is drying and Angus has spread tonnes of seaweed on his fields. He'll soon be ploughing.

I spoke to Donald MacKinnon at The Crofting Federation and he was impressed with the number of young crofters present who knew what they were talking about. NatureScot announced compensation of £6000 for 2022 and 2023 for North and South Uist for the goose problem. It used to be £40,000. Dare I say that they would be better giving the money to Ukraine! When I was Chairperson of the Crofting Federation and attended umpteen meetings in Edinburgh concerning geese, I was a lone voice and was used to hearing, 'We're not interested in reducing goose numbers but we do want to protect your crops.' Of course I always answered that you can't have the one without the other. Now they're not even interested in the crops. Anyone with a gun will have to shoot the geese when they're on their nests, when they're moulting and can't fly. It might be illegal, but allowing geese to destroy our crops and grazing should be as well.

There has been a war in Ukraine for four weeks now. It's heartbreaking to watch the news. The Ukrainians are strong and don't want to be ruled by Putin whom they see as a dictator. Poland and other neighbouring countries have been generous towards the poor refugees. Britain has sent clothes and blankets, as well as anti-tank weapons and lots of money to pay soldiers' wages. I hope that refugees from Ukraine will come to the Western Isles.

I have chosen this time, for your monthly texts, Psalm 84 and Psalm 107.

May

There's so much talk these days about self sufficiency, climate change and global warming, words I don't remember hearing when I was young.

Much of the coastal erosion is because causeways were built without sufficient sea openings, like Baleshare, and that has an effect on cutting peats. When peats were cut by hand, most people put the turf grass side up and closed up the area where the peat had been extracted. They made sure that the water trapped in the ditch drained away by digging out the lowest end. Now the ditches are blocked making the land unsafe for livestock. Of course, there's money involved, but all I know is that landowners are paid for locking up the carbon on their land. It makes me mad to think what our forebears did to dry up the bogs so that they didn't become quagmires.

After the second world war, the government helped farmers and crofters to cultivate their land so that there was enough food for everyone. The Department of Agriculture gave crofters a tractor and a driver. I can remember excitedly going down to the machair after school to get a ride. I can't remember if we had to pay or not. During WWII there was a *Dig for Victory* campaign to encourage people to grow their own food whilst rationing was in place. Today, with rising food prices, it might be a good idea for everyone to grow for themselves, in pots or in a little piece of their garden. Tagsa Uibhist Community Gardens at East Camp in Balivanich are running a *Generations Gardening Together* project and everyone is welcome – a great place to pick up tips! I worry that trees are being planted instead of crops.

The covering of beautiful, flat, fertile land is a huge mistake.

The war in Ukraine is dreadful, hundreds of civilians killed and tortured. It's terrible for people to see their towns, villages and cities being flattened. The Ukrainians are strong, and are determined to keep the Russians out.

My recommended readings for this month are, Mark, Chapter 2 and Psalm 34.

June

The local election is now over and those elected have a lot of work ahead. I've never heard so much party politics discussed. I was happy to see that our two stood as independents. They recognised that because their constituents are from different parties so surely the councillors should be too.

John and Brenda come to Kyles every year and stay in the Bramble's house. Over the years I've got to know them really well and every year they bring me something nice. This time it was the most beautiful varnished walking stick which John made. The leg is of hazel with three natural twists and the hand grip is beech, shaped like a goose head. His signature is in the wood and a silver stag's head is below the handle grip. It is a piece of art, attractive and unusual. When people come to visit me they see the stick before they see me!

On the first page of the Scottish Farmer was a story about a sea eagle. A farmer on Mull witnessed an attack on one of his lambs. First, he saw two peregrines attacking a sea eagle. He admired the eagle's defensive skills as it turned on its back and used its talons to defend its body. Then the peregrines gave up and the eagle flew away. But then the eagle dived below the cliff and when it reappeared, it had a healthy lamb in its talons. The farmer could hear the lamb crying. The eagle then landed to kill the lamb. The farmer ran towards it with his dogs, startling it and making it fly fifty metres away. They continued chasing it and when it landed the second time and spotted the farmer, it was distracted and the poor lamb, or lucky lamb, wriggled free, and the eagle flew away without its dinner. The lamb's mother, who was nearby and crying, rushed to it for a very happy reunion. The lamb had a gash on its neck but it was not life threatening.

It was cruel of SNH to reintroduce these birds. Their first reintroduction in Rhum was in the 70s and they came to Uist by themselves. For ages, SNH tried to convince crofters and farmers that sea eagles would only take dead lambs, but now they will have to admit that they take healthy and live ones too. On Vallay, three calves have been taken over the years, two white and one red. Shepherds have lost fifty lambs in a year. You read about the management of sea eagles and compensation for losses, and some shepherds have even used these funds to build sheds for lambing. No doubt, in a few years, when the sea eagle population has increased, NatureScot will stop all grants and concentrate on another species by which time some crofters will have given up sheep altogether.

Calving is doing well, though slowing down with about twenty left to calve. There have been two sets of twins, one from the first pair unfortunately dead, but the other set, one male, one female, is fine, and the mother has plenty of milk and loves them. Very often the female twin won't breed. We call them freemartins. Anyway, she will be kept and get a chance to breed. Once calving starts, Angus leaves a cattle trailer over on Vallay so that if anything is wrong, a cow can be brought back quickly. When Carianne was over, she found two cows who had mixed up their calves. She always knows when there is a mistake because she tags the calves when they are born and each tag has a unique number. Shona was the quietest one so Carianne put her and her calf into a small field while she attached the trailer behind the pick-up. She then had to catch the calf, put it into the trailer so that Shona would follow it in, and then drive both back to Kyles. Michelle calls her the Cow Whisperer because the cows trust her and she has a good way with the animals. When she reached Kyles, she got the younger ones to help her put Shona in the crush at the pens so the correct calf would suck its mother. It took a few days to get mother and calf reunited properly by putting them in the crush twice a day. Carianne and Ryan took the other cow and calf home and the same had to be done for them. We still don't know how the mix up happened but both cows and calves are perfectly matched and we are proud of Carianne. She noticed the problem and got the quiet cow home first all on her own, knowing that the men were busy and that the cow trusted her.

I hope that the weather will get better soon and that the crops will have a good yield. The Psalms that I choose are 80 and 84 for this month's reading.

October

I remember the coronation of our late Elizabeth II in 1953. Although she reached ninety-three years, her death was still a shock and so sad. I was pleased she was in Balmoral, in her beloved castle. In Uist, there wasn't electricity until the late 1960s and few people had a radio so we didn't know what was going on in London at the time of the coronation. However, our neighbour got newspapers from the mainland and we saw the news and lovely photos.

We've had a good spell of weather. My garden is good but the weeds fairly grew back when the sun was shining. The plants are better than they've been for years. We've eaten all the cauliflower because once ripe it won't keep. Lettuce, beetroot, onions, cabbage, carrots and kale are looking good so there will be plenty of nutrition for the winter.

Jean and I have made stovies twice this year so I thought I

would let readers who haven't made them know how to make this simple dish. I have a stove but a slow cooker is the nearest thing if you don't have a stove, and this must be cooked very slowly!

Use a large pot, about 4 litres.

Add one ounce of butter.

Chop 3 large onions (everything must be chopped small) to cover the bottom of the

pot.

Add potatoes, carrot, turnip, cabbage, lettuce, broccoli and kale.

The only thing I add after that is salt.

Stovies are tasty because water isn't added and the vegetables cook in their own juice. Some people cook or serve meat with them but I never do probably because my mother never did. I can still remember how we could smell the stovies when we came home from school.

Once upon a time, Scotland was full of hard-working young people but there seems to be a shortage of workers all over the islands, the same on the mainland. The Scottish Farmer reports that fruit growing farms are really short of workers. Some of this is caused by Brexit, but it is also reported that Scottish workers want to drive machinery and won't dirty their hands. Some of you will remember the Job Creation Scheme in the 70s when townships would create work like cleaning rivers, building cattle pens and sheep fanks[1]. They didn't get big wages, but enough to pay the bills until they got something better.

It was nice to hear that our late Majesty the Queen liked Psalms very much and that her favourite one was Psalm 23 so I think that I will choose it again for this month and our new Prime Minister chose Eccelsiastes Chapter 3, Verses 1-8 which I have favoured before so I will leave you to enjoy those.

November

On October 1st, eighteen young women and seventeen tractors set off on a long journey from Eriskay to Berneray to raise money for cancer relief in the Uists. They got breakfast in Eriskay, lunch in the Dark Island, tea with home baking in Carnish Church Hall, and dinner in Berneray. They were well fed by the local community. They collected £18,800, some achievement.

Angus had four tractors taking part and the week before he had to train the drivers. They were good car drivers but those new tractors are very different. He was teased when locals saw him every evening with different young ladies in his tractors! The ladies drove in convoy at the same speed. I went with my cousin, Euna, to Bayhead shop to give them a wave. I thought they would appreciate a crofting granny wishing them good fortune. I want to mention the strength and determination of one driver, Chrissie Laing who has motor neurone disease. There she was smiling with the rest of the team. As they drove past, I had tears in my eyes, nostalgic, but also proud of each of those girls.

Calum Iain Mhor, Calum MacRory, lived in Balemore. People in their 70s will remember him and those younger will have heard of him. He was a crofter and Christian. He had a powerful voice, adored children, and they visited him and walked beside him when he was ploughing with his horses. When the children played near his house, he would be singing hymns and psalms and be praying on his own. He was a bachelor and lived alone but he visited an old lady regularly, and told her about an experience when he was working at the corn on the machair. He heard a deep voice. There was no-one there, but he knew it was the voice of God. He said he saw a sign in the sky but no-one alive knows what it was.

The Psalm for this month has got to be 121, 'I to the hills ....'and Isaiah 53.

December

I wonder if anyone responded to the idea of a tunnel across to the mainland? I'm thinking that there wouldn't be much support for that. As I write, I'm watching the south-easterly wind driving the waves furiously towards my house. Imagine going underneath those waves! Nor would it be safe to have our islands open to anyone at any time of the day or night. It would be better to support the ferry service. Although it's often broken or stranded due to bad weather, there may be better ferries in the future.

I don't know when Lyme Disease arrived here. I was thirty before I saw my first deer on this land. It wasn't fear of humans that kept them out on the moor, they stayed there because they had enough green grass. The sheep and cows were summered out on the hill until October and the moors were methodically burnt and so with the sheep, lambs, cows, calves and the deer grazing there, the management of the moor was healthy.

During the war when the men got home on leave, the deer were a Godsend but they had to be careful which ones they ate. I have already written about my Uncle Hugh shooting a deer. I was too young to eat meat, but the others ate it. Possibly the deer had been wounded and its meat was poisonous. Certainly all the other children were very sick and it was only Doctor AJ's skill that saved them.

The sheep didn't have so many ticks either because the burning of the muir killed a lot of them. However as time went on, the burning was not done so methodically so the heather was able to grow. Then in 1961, reseeding became popular and the gearraidh² areas became greener, attracting deer, and the moor became rank through lack of grazing. Muirburn should be done in different areas every year with plenty volunteers to help although we have trees now so they would have to take care not to burn them. Deer numbers should be very much reduced but a huge cull would be heartbreaking. I'd like to see hardy sheep grazing on the moors along with the deer but that will only happen if crofters get Hebrideans and Blackface because after muirburn, there is no tall heather. The government should finance those sheep becoming hefted on the moors which means that they would settle and stay there. In the late 70s, we burnt the ranch, as we called it, on Marrival, and when the grass grew the following year, it was lovely and green. The cattle loved it and were able to summer there until 1997 when the fence stopped being stock proof.

In early November, the young ones were very busy in Vallay checking the cattle and giving them their medicines before the winter. It takes a lot of planning when the days are so short and work is governed by the tide. On Friday 11th, the tide was completely out at 4 pm so they took everything over to Vallay to use the next day. They then gathered the cows and their calves and put them in the field around the old houses. Saturday was the big day. Carianne, Fraser, Alexander, John MacPhee (the Wee Man), Craig, Ryan, Alasdair Don and the scanner, another Ryan, and Hector Shepherd took them over in a boat at 7 am giving the team all the daylight hours to complete the job. The cows were scanned for pregnancy and were treated for worms, pour-on for lice and mineral boluses. Since we have given them those, there's been no white scour in the calves. The day went very well. Two cows, Morag Skye and Sobhrach, with twins from springtime, were brought back to Kyles to be looked after. It's a big job with one hundred-and-thirty-nine cows. Carianne knows each one's name, takes a note of the ear tags and pregnancy information. She's a real cowgirl!

Our most exciting news that day was that Big Boy's son, Little Big Boy, had managed to impregnate all the heifers that had been put with him. He had been privately sold, but on inspection, before he was due to leave, we saw that his long hair had been caught around the end of his most important tool and had to be cut off. We couldn't sell him but wanted to give him a chance to heal and perform. He was put with heifers and kept in Kyles. Well, he didn't let us down and we look forward to seeing his calves next spring!

We are almost at the New Year and this is the last proper paper of 2022. I hope and pray that God will make 2023 a happier and peaceful year. We must be thankful for all the joy we've had whilst remembering those no longer with us. My choice of readings are Matthew, Chapter 1 and Psalm 118.

Seasons Greetings to you all!!

2023

February

When I was young I loved the snow and frost. I looked forward to it. We haven't had heavy snow since 2011 so this heavy fall is a change. The torrential rain that we had in December, in fact for most of 2022, has left the ground very wet and now that it's been covered in melted snow, it's hard for the livestock to find a dry bed. Maybe the summer months will be warmer than usual. I, for one, will look forward to that!

It was Angus's birthday on the 17th January and Michelle and I were reminiscing. He was born in 1965. There was heavy snow then too and we didn't have the comforts we have now. My husband was in Australia, but I had come home to North Uist to be a bridesmaid to my sister, Jessie, who married David MacKenzie in Glasgow in August 1964. I was carrying Angus and the local doctor, Dr A J MacLeod, advised me to stay here rather than travel for a month by boat back to Australia. Angus was due on Thursday 22nd January, but he decided to come and see this world on the 17th. Right from the start, he was the boss and liked to be early at whatever he had to do. At 6 am on the 17th I guess that his patience had run out. I woke my parents. My mother had five children so she knew about childbirth. There was one phone in Kyles village, a kiosk in Seonaidh Dhottaidh's house, but it was out of order probably because of the snow and no-one had a car. My father had to walk to Bayhead to get Domhnall Uilleam who worked for the water board and had a vehicle that was able to take my father to the nurse in Hougharry. So then my father and Nurse Katie MacDonald came back to Kyles in her car. She then took me to Lochmaddy where Dr MacLeod examined me and decided to send me to the Sacred Heart Hospital in Daliburgh, South Uist. Now as most North Uist people have heard, Dr MacLeod was uniquely skilled in obstetrics. While waiting for a taxi, he told me that there was a slight problem with the baby's head position, but patted me on the shoulder and said, 'You'll be okay, you're a strong girl.' In those days the roads were bumpy and more winding. I didn't feel the journey was too long nor was I nervous. In Daliburgh,

I was examined by Dr A MacLean from Raasay who was all smiles and said, 'The bumpy journey didn't do you any harm, in fact it did you a lot of good. The baby's head position is perfect now.' I was very happy to hear that. When I think back, I know that God was looking after me the whole time, just as he has my whole life. Angus was born at five minutes to midnight. Being in the Sacred Heart Hospital was a lovely experience. The nurses were mostly nuns. I can only remember one nun named Justina, and one nurse, Rhona Lightfoot, who was a great piper, I believe. They kept me in for ten days and I was so happy there that I cried when I left.

Where the Tractor Shed is today on Kyles Road, a knitwear factory was built and opened in 1970. The factory was a busy place with plenty of jobs, mainly for women. It belonged to MacKinnons of Scotland Woollen Mills, Coatbridge. Jumpers and cardigans were knitted and sewn up to the armpits with machines. The collar was knitted separately. Everything was folded up and posted to the Coatbridge factory once a week for finishing. There were about twenty working there with the workforce coming from all over North Uist, some from Benbecula. The Benbecula women were mostly army wives and some would only last a few months because their husbands would be transferred and they would leave the island. The first manager was the late Lachlan MacDonald, but he only stayed a few months. Then Angus Shepherd came and stayed for eight years. Angus always helped the women to mend their machines if one got broken. Dolina MacDonald was manager when Angus left. Effie Ann MacPherson was one of the main knitters, and she drove the van that picked up workers on the west side of North Uist. Effie Ann worked there from 1970 to 1980 as did Annie MacDonald. I worked in Bayhead shop until 1974 and the knitwear girls were great customers. The factory closed in 1980 after MacKinnons lost their contract with Marks and Spencer with only a week's notice and everyone was out of a job. The building was eventually demolished by my son, Angus. We used the walls, wood and corrugated iron for shuttering the poured concrete for our first decent shed.

March
Dùthchas
It must have been a treat for the Berneray community to see the film Dùthchas. It would have given them mixed feelings, some joyous, some sad. Ann and Bill Scott had been filming since the early 60s and that was long before the causeway. They died in the 80s and the films they had made were found recently in a box. Andy MacKinnon at Taigh Chearsabhagh was one of the directors who put the final film together from Ann and Bill's footage. It was shown first in Carinish and Berneray halls and has now been shown on BBC Alba where I saw it. It was a silent film and of course the Berneray folk were much younger. I loved the part where Gloria and Donald Alick MacKillop got married. It was 31st May 1968 and so windy that Gloria's veil kept flying straight up. They looked so young and happy.

I was speaking to Gloria recently. She is now ninety-one years old and had just been to Australia to see family and friends. She travelled alone by plane, there and back. What an amazing woman! The late Donald Alick, 'Splash',was ever so cheery too. Gloria was telling me that when Prince Charles visited Berneray and stayed with them, he helped them dipping sheep. When they dipped the last sheep, a posh voice said, 'Splash should now get dipped.' The Prince adored Berneray and had some of the most relaxing days in his whole life.

Oban Sale
The Highland Cattle Sale in Oban was on 13th February. We had intended to sell eight pedigree heifers and a black bull, but things were very busy at the new fish plant so it was decided that Angus and Fraser would need to stay here, so the plan changed. Carianne and Alexander would go on their own and take just the bull. Alexander is now seventeen and very capable

with the animals. Carianne, as most people here know, is a natural cowgirl. You can follow her on Instagram and Facebook – the Hebridean Cowgirl with her animal pictures and stories. It was a windy week with gales and rain so the only day that they could get away was the Thursday before the sale. They both drive and going via Lochboisdale to Oban meant a shorter road journey than Lochmaddy to Uig. They arrived in Oban at about 9 pm. I suppose Carianne was quite nervous as this was her first time away to Oban without Fraser and Angus. The other breeders normally arrived on the Saturday so Falaisg, our black bull, had peace and quiet to settle in. There were a number of Germans there and Mario, the Swiss lad who had worked here in 2021 with his father. He is due to come back to work here again in the summer and we look forward to seeing him then. At the show, Falaisg was not placed but this is only the judge's opinion. Sale day, not the show, is what we worry about. On Monday, the sale started at 10 am and the bull was number 24 in the show catalogue. The ferry home was due to leave Oban at 12.45 pm and ferries were unpredictable for the rest of the week because of the weather so we hoped that he would be sold and Carianne and Alexander would make it on time. I could see from the photos how handsome Falaisg looked whilst they were waiting their turn to get into the sale ring. He was lighter than the rest but really beautiful, his head and horns just perfect, his back straight and legs as they should be. In fact he was very correct. I can just imagine what was going through Alexander's young mind as he took the bull into the ring. Of course he has done that before, but his Dad and Fraser were always nearby. The Champion bull made 22,000 guineas. The next highest price was 6,200 guineas. Then number 24, Falaisg, was in the ring with Alexander holding the halter and Carianne standing below the auctioneer. I'm sure their hearts were thumping. The bidding was fast and in no time Raymond the auctioneer dropped the hammer at 6,000 guineas. Pedigree animals are still sold in guineas – one pound and one shilling. The nice thing about it all was that Falaisg was going to Mull. Most people here will know that our family has strong connections with Mull because my mother came from there. Because Carianne and Alexander had to catch the one and only ferry, they could not stay to watch the rest of the sale and had to rush off to catch the boat which sailed to Barra and then on to Lochboisdale. It was the daughter of the buyers who bid for our bull so no doubt Carianne found a minute to talk to her. She had visited the bull in his pen but had not let on that she was interested in buying him. I've seen the photograph of Alexander with Falaisg in the Scottish Farmer. He and Carianne did very well, and Angus and Fraser are proud of them.

Cattle Inspection

Tuesday 21st February 2023 was cattle inspection day, the first since Covid began. The Department of Agriculture can usually come whenever they want, but in our case, with the cattle being on a tidal island, we must plan the inspection days very carefully. The moon was new on Monday so the tide was suitable this week for getting to and from Vallay during the day. The cattle have to be moved from the north east field to near the pens the day before and silage must be spread on the ground on the monument side where they will go after the inspection.

Some of you won't know what is being inspected or why. Every animal has a passport issued by BCMS, the British Cattle Movement Service, giving the animal's date of birth, its tag number (the holding number and a unique number for the animal) and the tag numbers of the mother and the bull that sired it. The breed, the colour and the sex of the animal are also included. Carianne must report births, deaths and movements of animals within a certain time frame or there are penalties with money taken from your subsidy. The inspection involves checking that no animal has lost a tag and that the whole herd is in a healthy condition. As there are about 130 cows, they have to be penned so that the two tags can be seen. The stock bulls can be viewed in the field because they are very tame. The inspector also checks that you have the correct passport for every beast. You might think that we are joking when we say passports, but it is a tremendous amount of work and it has to be done. The inspection has now been completed and all our records are correct. Well done to the team – Angus, Fraser, Carianne, Alexander, Wee Man, Craig and Ryan.

April

At the beginning of the month, Jean took me to the Uist and Barra Hospital in Benbecula for an appointment. Thankfully all was well. In the waiting room, two ladies came over to me but it's difficult to recognise people when they wear masks.

'Are you Ena?' one said.

What a lovely surprise when she said, 'We're from Eriskay. Eunace and Isobel. Remember we used to be receptionists in the Royal Hotel where you always stayed when you went to the Highland Cattle sale.'

Of course I remembered. I was so happy to see them again. It all came back how they would always find me a room even when they were full. Isobel had left Oban twenty years ago so she did well to recognise me.

Then Jean and I got permission to visit Katie Wilson who had been in the hospital for a few weeks. It was great to see her looking so well and she was pleased to see us. She told me that she had enjoyed reading in Am Paipear about my stay in The Sacred Heart Hospital in 1965. I had mentioned a lovely nun, Justina, who worked there then.

Katie said, 'There's a lovely nurse here and she's called Justina. I gave her the Am Paipear and seemingly she is named after the Nun you wrote about. You need to see her.' Then who walked past but Nurse Justina. Apparently she was born in the Sacred Heart Hospital in 1970 and the nun, Justina, delivered her. The nurse's mother was so delighted with the nun's care that she decided to call her baby after her. Nurse Justina was ever so excited when Katie told her that I was Ena who wrote in the paper. We hugged each other and I don't know which one of us was more happy. It was one of those moments that you never forget.

It was wonderful to get a week of dry weather even though there was also snow and frost. Angus took advantage of the dry days and spread hundreds of tonnes of seaweed to fertilise the next crops. While I'm on the subject of weather, I must admit that I really enjoy listening to the weather forecast when presented by Derek MacIntosh. He makes it easy to follow with his fluent Uist Gaelic and always tells us what to expect. Recently he said that it was going to be very cold tonight so make sure you keep yourself warm. Of course, he comes from North Uist and no doubt he was thinking of his Granny. You are a good ambassador, Derek. Thank you for always being so thoughtful.

The calving and lambing will be starting very soon so let's hope that crofters get better weather. It's good to see that there are lots more native breeds of cattle in Uist. I love to see Aberdeen Angus and Shorthorn bulls all over the islands.

The war in Ukraine is still raging. How the Ukrainians manage to survive is beyond belief. I was pleased to hear on the news today that President Putin is at last accused of war crimes by the International Criminal Court in the Hague. It is all too complicated for me to understand but maybe this will be the beginning of the end. One thing that bothers me is why the West makes public the support they are going to give to Ukraine. You would think it should be a secret.

July

The last time we had weather like this was in 2011. Before this heatwave, we had dreadful rain and wind most days, and that rain meant that the ground was full of water. However, this recent good spell has left the ground a bit too dry and on the 10th June in the evening, there came a heavy shower which watered all the crops. Since then we have had a few more showers. A little and often would be a Godsend.

A few weeks ago, Angus took me in the pick-up to see the Paible machairs where we checked the nine heifers and nine calves. They all looked well but I know they were looking forward to grazing in the hill pastures and now they are in their favourite summer grazings with access to the lochs. We went from there to see the crops of oats, rye and barley and saw that none had suffered yet from the heatwave. Then we went over Knockline, Balemore, Knockintorran and Paiblesgarry and I was really happy to see the machairs still green, I guess a result of the winter's heavy rain and Angus's covering with seaweed.

I wonder how the corncrake numbers are this year? I have heard many of them and it could be that the numbers are up. I have heard the males calling but I haven't seen any yet. I was lying in bed around 6 am listening to the bird song. The corncrakes and doves seem to be competing and I heard the cuckoo twice. They were all so close to home. Then a light shower fell which was more exciting to me than the birds. There are two starlings nesting inside the frame of the wee shed in front of my scullery window. They have been there for a month and I watch them together a lot. The main problem is my cat. Luckily she has not noticed them. She is out all night, comes in in the early morning and sleeps most of the day. I feed her a lot so that she's not hungry, and to take her mind off the birds. Jamie Boyle, RSPB, tells me that starlings sometimes have two broods so that's maybe why they have been around so long, although I haven't seen any chicks yet. Cissy, the cat, did catch a sparrow this evening. I think her teeth are getting bad because she left it in the grass. I thought that it was dead, probably through shock. I picked it up and I could feel a movement so I put it inside a dish and very quickly I could see that it was getting stronger. Angus was over and he took it outside. As soon as he took the lid off the dish, it flew away,

275

towards a house that doesn't have cats, so that was a good rescue for the wee bird!

On Sunday 11th June, we had a beautiful Gaelic service in Taigh Sgire, Sollas, led by Reverend Ronnie Morrison. It's hard to believe how things have changed. When I was young, all the church services were in Gaelic, but if there was someone in church who didn't speak Gaelic, the Minister would give them a summary in English at the end of the service. Naturally this extension to the already long service didn't go down well with us young adherents! Gaelic speakers looked forward to this recent service, probably about forty attended, and Reverend Ronnie told us that it was fifteen years since he had last held a Gaelic service in Dunskellar. After the introduction, two young girls, Kate Ellwood and Rebecca MacDonald, read Luke, Chapter 8 and Mark, Chapter 4, verse 41. The singing of the popular Psalms 100 and 107 was uplifting. The presenters were John MacAulay, Angus MacDougall and John Angus MacKillop, whilst Fergus MacBain offered up a prayer for the congregation. Reverend Ronnie took his sermon from Mark, Chapter 4, verse 41. Mark and Luke wrote the same story, but Mark has more detail in his telling of it. Although I have read both often during my life, I must admit I can't remember the sermons. I never understood how the storm, which came by surprise on the Disciples' boat, did not touch the small boats nearby. It's a wonderful story about Christ controlling the sea and the wind. Thank you, Reverend Ronnie.

August

I was fourteen years old when we got our first tractor so I was old enough to have helped my father with that work. The cart wheels were narrow and to protect the horses' feet the road needed a good foundation so we would cover the road in sand sods every year. Fortunately, we lived beside the sea so it was easy to cut sand sods.

One day Sally, our young mare, did get one of her hind legs stuck in the peat mud and we had a difficult job getting her out. Thank goodness that only happened once. Before the horse and cart were taken to the peats, we did everything to save time. We always went there when the tide was out so if it was the spring tide we would have approximately four hours a day to go backwards and forwards across the beach to Kyles with our loads. The peats were put in heaps ready to be thrown onto the cart and there was no way the horse could leave the sand sod road. If you were lucky to have your peat bog in a fairly flat area, you would have carried the peats in a wheelbarrow, but lots of areas were uneven and we used either a hessian sack on our backs or else a hand barrow which needed two people to lift it. It was just a square or rectangular piece of wood with two handles both back and front. You had to watch where you were going because a stumble could make you fall and lose all the peats! I can't remember how many loads we were able to get home in a day but not very many. There was no lack of 'hands' as mainland cousins would be here for the summer to help. My brother was at home until 1954 and I suppose my sisters who were all at college were at home for the summer holidays, except for the oldest, Flora, who was in Glasgow at college and working there too, so unable to come home. Every time that we brought a load of peats home, we would take a load of sods and sand back to the peat road with us which in those days was just gravel. I can't remember when it was tarmacked. When all the peats were home, they were made into a beautiful stack to keep them dry and tidy. This was often made by the older men who didn't do the heavy peat work early on. Remember families were larger then and one man might still live at home other than the crofter. My uncle Angus, who had asthma from an early age, was an expert at those jobs. When all the peat jobs were complete it was nice to know that you had enough fuel to keep the house, and all who lived there, warm all winter.

Morag MacBain and I were reminiscing recently about the people who lived in Bayhead when we were in the primary classes in Paible School. I thought it would be nice to mention where they lived. Some of the young will not have heard of those people so maybe Am Paipear is the best place to keep them alive. The houses that I am talking about were opposite Donald Ewen Nicholson's house and belonged to cottars, people who worked on the land but had no land of their own. They would have had other jobs to be able to survive and feed themselves. The houses all had thatched roofs and looked cosy and were built very close together. Here are their names: Flora MacDonald who was a school cleaner and daughter, Mary Ann; Archie Millar who was well over six foot, had tufts of unshaved hair on his cheeks and taught people the psalm tunes; Kate, the tailor's daughter who I remember cutting potatoes for us in Kyles before we planted them; Martha, the daughter of Iain the blacksmith who was also a school cleaner and Maggie MacDougall who had a daughter Margaret and a son Donald John who was born in 1926. Margaret worked in the school canteen but Donald John was very famous in our circles for what he achieved. Why he hasn't been written about before I do not know. Donald John left school at about fifteen and did various jobs. He worked on Kirkibost Island farm and then in Benbecula at the construction site of the aerodrome after which he joined the Air Force. He married Rosie MacDonald from Paiblesgarry and they had four children, Angus, John, Donald and Mairead. Donald John became a Squadron Leader and I can still remember him – a good looking man and especially handsome in his Air Force uniform. He flew the aeroplanes that refuelled the other planes up in the sky and that wasn't an easy task. He spent some time in Singapore but was too young, a teenager, to serve in the Second World War. Paiblesgarry crofters were gathering sheep one day and two aeroplanes flew low above them. They were very startled because the planes' lights were flashing as if drawing attention to themselves. Soon, one of the crofters realised that it must be Donald John and it was. Calum Iain was just a boy and he was along with the crofters gathering the sheep. He got a terrible fright having never seen a plane so close before. I have always admired what Donald John achieved from very humble beginnings through his own diligence, determination and hard work.

September

My four hens are great layers and are very tame. I got the Welsummer and Rhode Island Red from Fergus John. They are real pets now and if they find the door open, they make their way in! The other two brown ones are old but don't look it. One I got from Snooker when he gave up the hens, and the other one was hatched here along with five white ones. The last one died a year ago.

About ten days ago, I heard a screech, looked out the window and there was Ralph with a mouth full of feathers chasing the brown hen at top speed. Angus and Mario chased him and luckily the hen was very fast and got away. However, she disappeared near the big shed and we couldn't find her. We looked everywhere and after two days I was sure that she must have died. On the third day, Angus let the other three out in the morning, fed them and came back with the empty jug. After leaving to go back next-door, he was back right away saying, 'There's four hens out there' and sure enough, there she was. She must have been in a secluded spot somewhere after being terrified a few days before and strangely enough there wasn't a mark on her despite all the feathers blowing in the breeze. I was really pleased to see her again because I was sure that she was dead. I still love Ralph, he is a lovely puppy. He was reprimanded and I hope that he has learnt his lesson. The only thing I can think of is that the brown hen went over to his patch and he didn't want her there and reacted when she ran away from him. Thankfully there was no blood spilt!

On the day that the hen was found, Cissy was in as usual and I noticed there was a swelling on her cheek. I couldn't see what was causing it but I suspected an abscess on her tooth. If it was worse the next day, I would get her to the vet. I'm usually up about 7 am and Cissy is always in the window wanting in. However, the next day she wasn't there and I thought it very strange. Once again there was a search party. My cousin heard a cat give one cry in Sarah's stable but when I went in no cat answered my call. We looked everywhere again so I went to bed with a heavy heart. Next morning, again no Cissy at the window. Michelle had a look in the stable with Eilah and Emily, Claire's girls whom she was looking after that day, and they all heard a cry. Michelle and the girls went on a hunt for a ladder and some help and found Iain Anderson and Ryan working in Kyles. Iain climbed up into the loft of the stable to get her – because there was a cat up there – but she managed to get away through the broken vent. Maybe she has been using that way in when it is cold and wet outside, and felt that it was a safe place when feeling poorly. Experts say that cats hide away when they feel ill so maybe that was it. Michelle and the girls then came to tell me that they had found her and lost her again! We saw her heading into the big shed so I put on my shoes and went out with them. We were hearing miaows but couldn't find her. We finally spotted her heading back out of the shed. Everyone held back whilst I went towards her, quietly calling her name and she followed me into the house. What a relief! Yet she looked so poorly and the swelling in her cheek was worse and her eye could hardly be seen. I took her to the vet next day and Emily the Vet discovered that her teeth and her eye were okay but the infection was from something sharp that had made a little hole below her cheek. She cleared out the poison and gave her antibiotics and now after plenty of TLC she is back in the window every morning. What a week for lost pets!

Young Emily said to me, 'What next is going to be lost? Are you going to be lost?'

I replied that I hoped not!

2024

February

This month, Ena MacDonald has been tied up with preparations for her new book. While we eagerly await its publication, Ena has chosen an item from her archives to keep readers busy until the next edition. The letter was written in 1980 in response to a call for contributions from the Scottish Farmer magazine which had offered a £5 prize for the best submission. It will be of no surprise to anyone who has read this column over the years, that Ena's letter was the one chosen for publication. We print it here with her permission.

Kyles,
Bayhead,
Lochmaddy,
North Uist,
10th November, 1980

Scottish Farmer,
"Is farming really such a thankless industry?"

Dear Sir,
As I am a crofter and not a farmer, I might not be entitled to write, however I feel I must write and tell the general public that farming, in my mind, is very rewarding. Our croft is situated on the windswept Atlantic coast where the weather has a lot to do with our survival. The city visitor will moan "what a wind" but to us, the strong North Westerly is the kind wind that will blow in the seaweed that we use for fertiliser. Great stuff this seaweed – produced (with a little help from cattle manure) the most tasty potatoes – oh yes I'm sure in the whole world. The barley oats and hay crop are also more palatable to the animals. The highland cattle adore the fresh seaweed and will often leave the green pastures for a 'salty lick.' This spring was like an Indian Summer and lots of farmers worried about whether their crops would grow – early June we were all tense but then mid-June came some rain, how beautiful it felt about your face, and the crops

seemed to grow overnight. July was perfect for growing and nature made up for her earlier laziness. If the weather was always perfect we just couldn't feel so overwhelmed with joy when the right weather did come. This year annual prices were so low and us crofters feel very disappointed but when you love your land you don't give up because the land would be bare without stock. How dull life would be without that newly born calf suckling its mother or that frolicking lamb that yesterday you brought in from the snowy blast seemingly breathing its last breath. Of course there's also these large pullets scratching around, did they really grow from these balls of golden fluff that I reared? Yes they did and today another two dropped their first eggs. Money is not easy to come by but then that's such a dull subject, next year things will be better.

Ena MacDonald

Glossary of terms and Gaelic words when first used in the text

Gaelic words are italicised. Where a Gaelic word is explained in the text by Ena, it is not repeated here, like '*suidheachan* (stooks of six sheaves).' Nor is machair italicised because it occurs so frequently.

2006

1. Shieling is a rough, sometimes temporary, hut or shelter used by people tending cattle on high or remote ground or pasture land for the grazing of cattle in summer.
2. BSE. Bovine Spongiform Encephalopathy.
3. OCDS. The Open Contracting Data Standard.
4. OTMS. The Over Thirty Months Scheme, to keep older cattle out of the human food chain.
5. SCF. Scottish Crofting Federation.
6. NGMRG. National Goose Management Review Group.
7. SNH. Scottish Natural Heritage.
8. In-bye. That part of the farm which is used mainly for arable and grassland production and which is not hill and rough grazings. 'In-bye' land has fields that are bounded by a fence, a dyke or a hedge. 'In-bye grassland' will be conserved for winter feed.
9. Machair. Gaelic word for fertile low-lying grassy plain, one of the rarest habitats in Europe which is found on exposed western coasts of Scotland and Ireland, and in particular the Outer Hebrides.
10. RSPB. Royal Society for the Protection of Birds.
11. Ardbhan. The home of the MacDonald family in Kyles, Paible.

2007

1. Stirk. a yearling bullock or heifer.
2. Og. Gaelic meaning junior or young.
3. SCVO. Scottish Council for Voluntary Organisations.
4. Crowdie. A soft, fresh cheese made from cows' milk, traditionally from Scotland.
5. Clegs. Relatively large insects with a vicious bite concentrated in the northern Highlands.
6. Fuaran. A well, spring or fountain.
7. Committee Road. Ena often references this road which is a narrow road which cuts across the desolate moorland in the NW of North Uist for 4 miles (6.5 km), connecting Claddach Kyles with Malaclate.

2008

1. Am Paipear is the local newspaper in which Ena's articles were first published.
2. Shaw. The parts of a potato plant that appear above the ground.
3. LFASS. The Less Favoured Area Support Scheme.
4. Blue Tongue. An insect-borne viral disease of sheep (transmissible with less serious effects to cattle and goats).

2009

1. Stooks. Stooks are an arrangement of corn sheaves usually six.
2. Toitean. Corn ricks.

2010

1. Muirburning.
2. ESA. Environmentally Sensitive Area.
3. RSS. Rural Stewardship Scheme.

2012

1. NFU. National Farmers Union.
2. Taigh Chearsabhagh. Museum and Arts Centre (and cafe, shop and post office) that champions heritage, visual arts and the Gaelic language and culture in Lochmaddy.

2013

1. Oor Wullie and the Broons. A Scottish comic strip published in the The Sunday Post.
2. Rylock. Galvanised wire for fencing.
3. Eorpa. Long-running Gaelic current affairs programme broadcast on BBC Alba.
4. Gaidhlig. Gaelic.

2014

1. Mee Beag. Mee is the sheep's name and Beag is Gaelic for small or young.
2. Mee Mor. Mee is the sheep's name and Mor is Gaelic for big or older.
3. Saithe. Pollock.
4. Cuddies. Small fish the size of sardines.
5. Sgoil Lionacleit. Six year comprehensive school on the island of Benbecula.

2015

1. Paible Junior Secondary. The old junior secondary school built in 1904.
2. Wedder. Scots word for a castrated male sheep.
3. Fatstock. Livestock fattened for slaughter.
4. Scunnered. Scots word meaning fed up and disgusted, often with a surfeit of food.

2016

1. Artic. Short for articulated.
2. Lairage. Pens, yards and other holding areas used for accommodating animals in order to give them necessary attention – water, feed, rest – before they are moved on.
3. Maragan. Gaelic for black and white puddings.
4. Objective One. Regional funding through the European Regional Development Fund (ERDF), the European Social Fund. (ESF), the European Agricultural Fund for Rural Development (EAFRD)

5. Feis Tir an Eorna. Gaelic school fair and festival held in July.
6. Chanter. An instrument related to the bagpipes and taught in preparation for them.
7. Shinty. A team sport played with wooden sticks and a ball.
8. Comhairle nan Eilean Siar. The Western Isles Council.
9. Sgoil Uibhist a Tuath. The new Primary School offering English and Gaelic Medium Education.
10. Cottar. In the Western Isles, someone without land who helps in return for food and other necessities such as fuel or a place to keep a house cow.
11. Saithe. Also called Coley.

2017
1. The Brahan Seer was gifted with "the sight" – an ability to see visions that came unbidden day or night. His prophecies were so impressive that they are still quoted to this day.
2. Toradh. Festival of food and writing.
3. Growthy. Exceptionally fast in growing and gaining weight

2019
1. Ospadal Uibhist agus Bharraigh. Hospital on the island of Benbecula.

2020
1. Air an Lot. Gaelic television programme.
2. Stovies. Traditional dish of vegetables made without water on the stove.

2021
1. Sarking. The material on a roof beneath the tiles. Sheathing, sheeting or decking.
2. Mamaidh. Gaelic for mother.

2022
1. Fanks. A sheep pen.
2. Gearraidh. Means rough ground separate from the in-bye. Often used in place-names in Uist such as Hogha Gearraidh (Hougharry) and Tigh a' Ghearraidh (Tigharry).

Milton Keynes UK
Ingram Content Group UK Ltd.
UKHW022040020524
442065UK00009B/88